Unl

The Colour of Metal Compounds

The Colour of Metal Compounds

Adam Bartecki

University of Technology, Wrocław, Poland

and

John Burgess

University of Leicester, UK

GORDON AND BREACH SCIENCE PUBLISHERS
Australia • Canada • France • Germany • India • Japan • Luxembourg
Malaysia • The Netherlands • Russia • Singapore • Switzerland

Original version published in Polish in 1993 as BARWA ZIĄZKÓW METALI by Oficyna Wydawnicza Politechniki Wrocławskiej, Wrocław.
© 1993 Oficyna Wydawnicza Politechniki Wrocławskiej, Wrocław.

Amsteldijk 166
1st Floor
1079 LH Amsterdam
The Netherlands

British Library Cataloguing in Publication Data

Bartecki, Adam
  The colour of metal compounds
  1. Metals – Coloring 2. Inorganic compounds 3. Colorimetry
  I. Title II. Burgess, John, 1936–
  546.3

ISBN: 90-5699-250-3

The authors are greatly indebted and most grateful to Mrs Mira Aleksiun-Żelechower for her gift of her original painting for the front cover of this book.

CONTENTS

PREFACE TO THE ENGLISH EDITION

The purpose, aims, and content of this English version are very similar to those of the original Polish book. There has, of course, been some updating, and we have corrected some minor errors – and probably introduced a few new errors in the processes of translation and editing. We have added some new material on solvatochromism, on colour in the diagnosis of mechanism, and on colour in teaching and textbooks, in all cases relating mainly to compounds and complexes of the d-block elements. Both authors of this second edition learned a great deal at the First Conference on Colour in Nature, Science, and Technology, held in Wrocław in November 1998 – some views and results presented at that conference have been incorporated into this book.

We are most grateful to Gordon and Breach Science Publishers for their enthusiastic support throughout the preparation of this English edition. AB wishes to thank Dr Głazek for invaluable assistance in the translation of the Polish original; we both thank Dr David Goodall for kindly reading the whole of the translation, for his helpful comments, and for his subsequent encouragement. We thank Professor Dirk Reinen for kindly providing pre-publication extracts from his 1999 Chemical Society Review which enabled us to produce Figure 7.18. JB would like to record here his long-standing gratitude to Dr and Mrs Grzedzielscy, who looked after him during his time in Ipswich and introduced him to Polish history, customs, and food, thereby laying the foundation for subsequent collaborations with several Polish institutions and thus to this present book. It is a great pleasure for him to acknowledge the generous hospitality of Professor Alina Samotus and her research group in the Jagiellonian University of Kraków, of Professor Narbutt and his colleagues at the Institute for Nuclear Chemistry and Technology at Warsaw, and of course of his coauthor at the Technical University of Wrocław.

PREFACE TO THE POLISH EDITION

The idea of writing a book on the colour of chemical compounds (of metals) was born ten years ago, when the present author (with a few co-workers) started his studies on quantitative measurements of colour and, generally, on the problem of the chromaticity of such compounds.

Colour plays an extremely important role not only in our everyday life but also in industry and art. It can, as a matter of fact, be a source of aesthetic feelings, but at the same time it provides us with valuable information about the structure and properties of material (coloured) objects.

A reasonable use of colour as a source of structural and analytical information on chemical compounds must be based on a deep knowledge of the interrelation between the colour and the properties of those systems. This requirement is not easy to fulfil as, apart from subjective aspects due to individual colour perception, the qualitative and quantitative features of colour depend on specific factors and measuring circumstances.

In chemistry, colour is a very important property of substances and because it is immediately observable, is given a concrete name. Since nearly all chemical elements can form coloured compounds, the use of colour as an analytical tool has a general significance. Moreover, the number of colour hues which can be differentiated is about 5.10^6–10^7. Hence, only quantitative measurements of colour can provide valuable information, particularly if the system under study undergoes some changes, for instance with time. Such measurements can be obtained by applying so-called trichromatic colorimetry, which is the basis for the CIE 1931 method. CIE stands for the International Commission on Lighting (originally in French – Commission Internationale de l'Eclairage). This approach (and some of its improvements) have since been applied in many branches of industry, e.g. for preparing pigments, in colour TV, in the technology of coloured glass. Applications in chemistry are practically unknown, with some exceptions connected with complexometric titrations.

Now, let us remember that hundreds or even thousands of coloured compounds are known, particularly compounds of transition elements and to a lesser extent, of lanthanides. Quantitative measurements of colour, expressed by the so-called chromaticity coordinates and luminance values, x, y, and Y, are obtained from the evaluation of absorption (transmission) spectra, and in some cases also of diffuse reflectance spectra. The author and some of his co-workers have been studying electronic spectra of transition metal compounds for many years. Such studies combined additionally with colour measurements seemed to be a purposeful and possibly fruitful method of finding out the connection between spectroscopic parameters and chromaticity features. This line of reasoning has been verified firstly by extensive experimental research and simulation of spectra and colour, and secondly by a thorough analysis of data in the literature.

Although colour is primarily a sensation, the entire problem of colour and coloured objects involves many different disciplines, such as physics, chemistry, psychophysics, physiology, psychology, applied medical sciences and even philosophy. As a result, this subject has been addressed in numerous monographs, books, and articles. However, the present book is in all probability the first one devoted to trichromatic colorimetry and CIE methods in chemistry.

The interdisciplinary character of the problem, with so many areas of science involved, makes it difficult to get a uniform picture of colour science. Moreover, the problem of terminology is a reason for other difficulties not only when comparing different languages, but also when comparing colour names used in everyday life and those used in colour science. For this reason, as far as the terminology used in this book is concerned, we have decided to use first of all the names formulated by the CIE 1931 chromaticity diagram, based on x, y values, and additionally the original names as given in various publications.

The first two chapters describe some general problems, qualitative aspects, and briefly colour systematics (colour order systems). The main interest lay in the quantitative evaluation of colour by means of trichromatic colorimetry and the CIE methods, and Chapter 2 discusses some interrelations between electronic (absorption) spectra and chromaticity. The basic concepts concerning the application of chromaticity coordinates and their significance for the chemistry and spectroscopy of transition metals are described in detail in Chapter 3. The next chapter is devoted to the colour of lanthanide ions and compounds. As is well known, analytical chemistry widely uses and draws upon the colour of substances, and this topic is the content of Chapter 5, in which qualitative, quantitative, and some specific problems are tackled. Chapters 6–8 address the following themes: colour centres, colour in mineralogy, coloured mineral pigments, and coloured glass doped with metal ions. The last chapter discusses some interesting uses of colour as a teaching tool, not only at an introductory level, but also in more advanced studies.

As has already been stated, this book is intended to fill the gap between the technical application of colour to such systems as e.g. pigments and coloured glass and more theoretical considerations based on electronic spectroscopy, but using colour only in a descriptive way. The book is formally restricted to inorganic (metal) compounds and does not cover the colour of organic compounds, which was the first object that in the past enabled the identification of some dependencies between colour and structure. It was in that area that many empirical rules, as well as the terms chromophore, chromogen, and auxochrome, were formulated. For the interpretation and prediction of colour, molecular orbital theories were used quite widely, but quantitative measurements by trichromatic colorimetry have only been applied to some dyes.

Not all questions connected with the colour of inorganic metal compounds are discussed in detail, and there may be some omissions, more or less serious. In part the justification for this is the limitation of the length of the book imposed by the (Polish) editor.

I would like to express my thanks to Dr T. Tłaczała for reading and commenting on the (Polish) manuscript as well as for calculating the chromaticity coordinates. My thanks also go to Dr J. Myrczek, who was responsible for computer simulations and took part in many discussions on the topic. I would also like to thank M. Raczko (P.E.) for measuring all the spectra with the usual great care.

It is also my duty to thank all editors and authors for their kind permission to publish the numerous data and figures.

Last but not least, I am greatly indebted to the referees, Professor P. Hawranek and the late Professor R. Sołoniewicz for their hard work in reading and discussing this book. The responsibility for any errors is, of course, mine.

1. GENERAL ISSUES

1.1 THE NATURE AND PERCEPTION OF COLOUR

As is well known, we owe the explanation of the nature of colour to Newton, who in his famous experiment dispersed white light (daylight) by means of a glass prism. As a result of dispersion, a coloured image was obtained on a screen, where a series of coloured patches could be seen graduating from red (for the smallest refraction angle of the ray passing through the prism) to violet (for the greatest angle). Newton rather arbitrarily distinguished seven colours, perhaps by analogy with the seven-tone musical scale. Table 1.1 lists the names of colours adopted at that time along with their respective approximate wavelengths.

Table 1.1 Basic and complementary colours

Colour (according to Newton)	Wavelength range (approximate) (nm)	Complementary colour (according to Helmholtz)
Violet	360–415	yellow green
Indigo	415–444	yellow
Blue	444–487	orange
Green	487–540	purple
Yellow	540–590	indigo
Orange	590–690	blue
Red	690–830	blue green

It should be noted that various divisions of wavelength ranges are used in the literature and different names are given to them. The chief objective in doing that is a more accurate delimitation of intermediate and mixed colours; their names are formed by combining the names of the component colours. This issue will be discussed in more detail in the subsequent pages and one example of such a division of the white light spectrum is given in Table 1.2.

White light dispersion is but one aspect of the general problem of the nature and perception of colour. As we pass to the question of the colour of material objects, we note that the occurrence of colour requires:

1

a)   a source of light, such as sunlight, a tungsten lamp, etc;

b)   a coloured object, such as a chemical compound, a painted artifact, glass doped with appropriate chemical compounds, etc.;

c)   a perception mechanism, such as the eye and the brain, or a measuring device, e.g. a spectrophotometer.

Table 1.2 A more detailed division of the visible spectrum and colour names

Colour	Wavelength (nm)
Violet	400–440
Indigo	440–470
Sky blue	470–480
Blue	480–490
Blue green	490–495
Green	495–560
Green yellow	560–570
Yellow	570–575
Yellow orange	575–590
Orange	590–600
Orange red	600–620
Red	620–780

This last item indicates that the problem of colour is not, and cannot be, limited to a physical process, which in the case of a coloured object consists of the interaction between electromagnetic radiation and matter (as well as the result of such inter-action), but also involves a subjective component, connected with the inter-pretation of the colour stimulus in the human brain. While the eye and the brain are physical factors, the interpretation of colour is to do with perception, which to a certain extent is individual, although colorimetry uses the important notion of an average (standard) normal observer, which is discussed in this book.

From the point of view of the aims of this book, the most important factor is the object, that is to say the coloured metal compounds. Thus, we have to give some consideration to the result of interaction between visible radiation and matter. Different phenomena may be involved, such as transmission, absorption, diffusion, and reflection (external or internal). Light transmission is at a maximum if the object is colourless. In the case of solids, the principle sources of colour are reflection and absorption, and, to a certain extent in some circumstances, diffusion. If light is reflected from a smooth mirror-like surface, the angle of reflection is equal to the angle of incidence; reflection from a rough surface does not share this characteristic – here one speaks of diffuse reflection.

In chemistry, light absorption by solutions is frequently studied – absorption spectra are an important source of information about the structures of compounds and their interactions in solution. Absorption spectra are also studied for solids, but for these it is more usual to obtain and study reflectance spectra. If the incident light is only partially absorbed, the transmitted portion of the light imparts colour to the

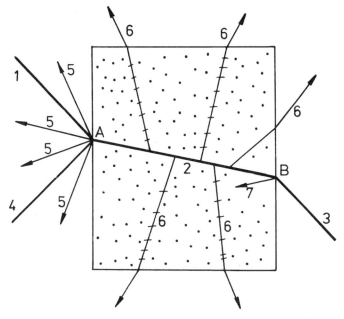

Figure 1.1 Scheme showing interaction of a ray of light (1), incident at A, with a translucent object. Light may be absorbed (2), transmitted (3), reflected (4 – specular; 5 – diffuse) emitted and scattered (6), or internally reflected (7).

solid, which is still transparent, or rather translucent. If all the light is absorbed the object is opaque. Figure 1.1 presents a diagram illustrating these phenomena.

Thus, it can be concluded that the interaction between an object and light may be described and analysed by absorption (transmission) or reflectance spectra. In principle, a qualitative assessment of such a spectrum enables an approximate prediction of the colour of an object to be made. However, especially if the spectrum is complicated, it may be difficult or impossible to predict colour.

Figure 1.2 shows the reflection curves for four different colours. The colour names are a qualitative description of colour perceptions, i.e. they are psychophysiological features. The generation of a colour stimulus, i.e. the psychophysical feature, is the product of three functions. These are the spectral distribution of the light source, the spectrum of the object, and the eye sensitivity curve [1].

Different white light sources are in practice characterised by different distribution curves, as indicated by the different spectral power distributions for the so-called A, B, C, and D_{65} illuminants shown in Figure 1.3. Standard illuminant A is an incandescent tungsten lamp operating at a colour temperature of 2854 K. Standard illuminants B and C consist of illuminant A shining through filters selected to simulate noon sunlight, while a series of illuminants D were developed to provide closer approximations to natural daylight (D_{65}).[1] More recently several standard fluorescent illuminants F have been established. The significant differences between

---

[1] The relation between these illuminants is indicated on colour Figure C1, q.v.

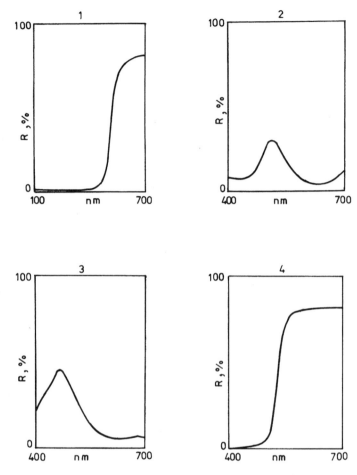

Figure 1.2   Reflectance spectra (schematic) and colour of opaque objects. R = reflectance; surface colours
1 – red, 2 – green, 3 – blue, 4 – yellow.

the various sources of 'white' light should be borne in mind. The various light
sources A, B, C, and F are physically constructed and their spectral distributions can
be determined and tailored experimentally – in contrast to natural daylight.

When considering the spectral distribution of eye sensitivity, one may find that in
real conditions a distinction can be made between the spectral curves for the normal
observer in scotopic and photopic vision. Photopic vision corresponds to a situation
where under the given circumstances (by day) there is strong radiation. Scotopic
vision is used at night. The photopic vision curve, which presents the values of the
relative spectral light response, is the basis of photometry and colorimetry. Its maximum corresponds to the wavelength of 555 nm [2].

The product of all the three functions is a function which corresponds to the
colour stimulus and, as will be discussed below, it is the basis of quantitative colour
calculations.

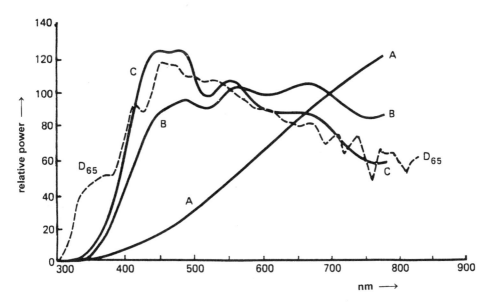

Figure 1.3   Relative spectral power distribution of standard illuminants A, B, C, and D$_{65}$ [6].

In a discussion of colour sensations, it should be stated that the perception of colour is an extremely complex problem for reasons which are not directly related to the scheme discussed above. It turns out that one should make a distinction between the perception of the so-called free colours and perception under real conditions, i.e. in a regular visual field. In the former case, other stimuli are irrelevant, while in the latter case, the sensation of colour is also connected with the perception of shape, size, and distance [2]. It has been found that chromatic colours and purple are perceived as free colours, while navy blue, grey, and olive are perceived as non-free colours.

1.2   BASIC ATTRIBUTES OF COLOUR AND THEIR DESCRIPTION

Different approaches can be used for the purpose of describing colour. One of them is the so-called 'desert island' (or thought experiment) approach [3, 4]. It is assumed that an observer who has previously not had any opportunity of observing and remembering a great number of different colours is staying on a desert island surrounded by an array of coloured objects. If this person is now faced with the task of segregating these objects according to their perceived colours, he or she will be able to solve the problem in different ways going through several discrete logical steps.

One solution is that the observer, having in mind the generally known names of colours (colour hues), first separates all objects with chromatic colours (those having hues) from those with achromatic ones (devoid of a hue), such as white, grey, and black. With a view to discriminating further the achromatic objects, the observer can rely on their lightness, which decreases from white through grey to black. In the Munsell system this feature is termed value (see Section 1.4 below).

Further differentiation of chromatic objects is more complicated. Of course, hue is a natural basis for colour distinctions, but lightness cannot play the same role as for achromatic specimens. There is, however, a third attribute – saturation – which can be different for objects with the same hue and lightness. This last feature indicates how much of the chromatic colour is mixed with the achromatic.

To sum up, we can conclude that this thought experiment creates a basis for a complete description of coloured objects. In other words, the observer may be certain that such a description is unequivocal. However, this set of three attributes corresponds only to psychophysiological free colours.

1.3   THE FORMAL DEFINITION OF COLOUR AND SOME TERMINOLOGICAL ISSUES

As mentioned in the Preface, the problem of colour is common to so many disciplines that a very general definition must not only be concise but must also present the most important feature(s). Such a definition reads: colour is a visual sensation caused by the visible region of electromagnetic radiation (by light of different wavelengths in the visible region). If we consider the basic features of colour, as described in Section 1.2 above, the definition would be as follows: 'colour is the attribute of visual experience that can be described as having the quantitatively specifiable dimensions of hue, saturation, and brightness (lightness)' [5].

Many specific terms are used in colour science and colorimetry. A glossary of some of these, including their definitions, follows [6].

1. Achromatic colour – colour devoid of hue, e.g. grey, white, black.

2. Achromatic stimulus – a stimulus that is chosen to provide a reference that is regarded as achromatic in colorimetry.

3. Additive mixing – addition of colour stimuli on the retina in such a way that they cannot be perceived individually.

4. Brightness – attribute of a visual sensation according to which an area appears to exhibit more or less light.

5. Chromatic colour – colour exhibiting hue (red, blue, etc.).

6. Chromaticity – property of a colour stimulus defined by its chromaticity coordinates.

7. Chromaticity coordinates – ratio of each of a set of tristimulus values to their sum.

8. Chromaticity diagram – a two-dimensional diagram in which points specified by chromaticity coordinates represent the chromaticities of colour stimuli.

9. Colour stimulus function – description of a colour stimulus by an absolute measure of a radiant quantity per small constant-width wavelength interval through the spectrum.

10. Colourfulness – attribute of a visual sensation according to which an area appears to exhibit more or less of its hue.

11. Complementary wavelength – wavelength of the monochromatic stimulus that, when additively mixed in suitable proportions with the colour stimulus considered, matches the specified achromatic stimulus.

12. Dominant wavelength – wavelength of the monochromatic stimulus that, when mixed in suitable proportions with the colour stimulus considered, matches the specified achromatic stimulus.

13. Hue – attribute of a visual sensation according to which an area appears to be similar to one, or to proportions of two, of the perceived colours, red, yellow, green, and blue.

14. Lightness – the brightness of an area judged relative to the brightness of a similarly illuminated area that appears to be white or highly transmitting.

15. Perceived colour – the visual sensation produced by light of different wavelengths throughout the visible region of the spectrum. By such perception an observer may distinguish differences between two objects of the same size, shape, and structure.

16. Saturation – the attribute of colour perception that expresses the degree of departure from the grey of the same lightness; or, alternatively, the colourfulness of an area judged in proportion to its brightness.[2]

17. Purity, excitation – the ratio of the distance on the CIE diagram between the achromatic point and the specimen-light point to the distance along the straight line from the achromatic point through the specimen-light point to the border.

Some other terms important for the content of this book are discussed in more detail in the respective chapters.

1.4 QUALITATIVE COLOUR SYSTEMATICS – OSTWALD AND MUNSELL SYSTEMS

One of the essential and basic problems in colour science and technology is to formulate a set of principal terms, internationally agreed and accepted, and to create more or less general colour systematics. Colour names themselves should be regarded as a natural system for the discrimination between the colour (colour hues) of material objects, e.g. of chemical compounds. If such a name corresponded to one single wavelength (single line) in the spectrum, no misunderstandings would ever be caused by its use. This is particularly important for mixed colours. In practice however, colour names describe some wavelength regions, which are not

---

[2] John Piper's baptistry window in Coventry Cathedral provides a good illustration of saturation. The reds, greens, and blues in the outermost sectors are strongly saturated, but the colours become progressively desaturated on moving towards the centre (W.D. Wright, *The Rays are not Coloured*, Adam Hilger, London, 1967, pp. 17–8 and Plate 7).

equidistant (as can be seen in Tables 1.1 and 1.2). Let us, for example, compare green and orange colours: the former comprises the region of 495–560 nm, the latter, merely 590–600 nm.

It can be assumed that Newton's well known colour wheel creates a kind of qualitative colour systematics or, more precisely, a system of colour hues. As expressed by McCamy [7], 'the hue circle is not an invented artifact' and 'the cyclic nature of hue is quite obvious.' One should also recall the colour scheme of Aristotle in the form of a sphere, the diameter of which contains chromatic colours, red, green, yellow, and blue, in addition to achromatic black and white.

In colour science, when describing and specifying colours, one speaks about colour-order systems and colour spaces. The latter term is related to the fact that at least three coordinates or colour attributes must be taken into account. The hue circle fulfils this condition only in part, but one more feature, saturation, can be added to this picture. If the centre of the circle is understood as neutral grey (achromatic light source), the distance from the centre to the circle corresponds to increasing saturation. However, the third attribute, brightness (luminance) cannot be shown within the plane of the circle. It may be depicted on an axis perpendicular to this plane.

There are many ways of classifying colour systems and spaces. One way is to differentiate between sets of coloured physical samples and collections which do not include such objects [4]. Within the first group, one class contains sets of coloured objects according to a concrete guiding principle; the objects of the other class do not share this feature.

Many collections are intimately connected with one branch of industry or another. In such cases, the choice of the required colour is restricted to a closed set of coloured specimens and intermediate hues cannot be obtained. There are, however, several industries (e.g. colourants for plastics and paints), where the final product is prepared by mixing the main colourant with white or black substances. This enables one to obtain intermediate colours according to the wishes or needs of the user.

Colour atlases have been created to ensure a perceptually logical and coherent system for describing colour [8]. In some cases such a collection is a tool for the classification of coloured objects according to their physical properties as well as to their colorimetric attributes.

To create more universal colour-order systems, one makes use of some quantitative approaches based on colour mixing laws. From this point of view, the most important and most objective is the CIE system, which will be described and discussed in the next section.

The Ostwald and Munsell systems [4, 6, 7, 9, 10, 11], now to be considered, apply to perceived colours, in contrast to the CIE system which is based on colour stimuli. Ostwald was a famous chemist (Nobel Prize, 1909); the basis of his system [12] is shown diagrammatically in Figure 1.4.

Experimentally, disk colorimetry is applied here. Colours are produced by mixing lights reflected from white, black, and highly saturated coloured segments of a spinning disk. Points W, B, and C in the figure denote the percentage content of white, black, and the so-called full colour, and, of course, $W + B + C = 100\%$. The

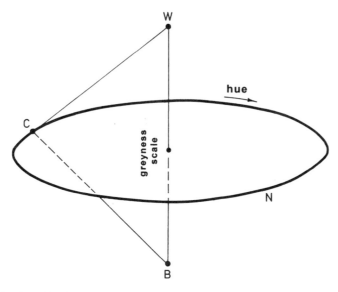

Figure 1.4 The Ostwald colour system [11]. The circle covers 24 full colours (C); B = black, W = white, and N = hue (see text).

letter N corresponds to one of the 24 hues which are based on four colours: blue, yellow, red, and green. These four colours are said to be opponent colours because an object cannot be simultaneously blue and yellow or green and red. The set of four colours, called 'tetrachromacy'[3] plays an important role in the theory of colour vision.

It is clear from Figure 1.4 that:

1. along the BW or black-white axis the grey content changes;

2. along the BC axis there is a change of the so-called shade;

3. along the WC axis different tints are formed;

4. by mixing the full colour with a specific grey value one gets various tones.

The resulting Ostwald colour series are summarised in Figure 1.5.

One of the most popular colour-order systems is the Munsell system invented at the beginning of this century [4, 6, 7, 10, 11]. Munsell was a painter who stated [10]

---

[3] Physiologically speaking, human vision is trichromatic – Section 1.5 of this chapter deals with trichromatic colorimetry – but a significant minority of people lack one of the three retinal photopigments. Red/green colour blindness is the most common form of this afflication. The resultant dichromacy, first studied and reported in a scientific manner by Dalton (J. Dalton, *Mem. Lit. Phil. Soc. Manchester*, **5**, 28 (1798)) is of considerable theoretical and scientific interest (see, e.g. H. Scheibner, *Vision Res.*, **38**, 3403 (1998); and H. Scheibner, in *Color Vision – Perspectives from Different Disciplines*, ed. W.G.K. Backhaus, R. Kligel, and J.S. Werner, de Gruyter, Berlin, 1998, Chapter 16) but outside the scope of the present book.

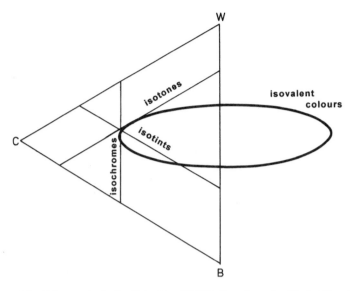

Figure 1.5   Colour series in the Ostwald solid [11]; C = colour, B = black, W = white.

that he intended to create a colour-order system applying measured physical quantities. However, in fact the system represents an order of perceived colours and is
based on the principle of equal visual perception. Munsell recognised the three
attributes of colour and named them **hue**, **value**, and **chroma**. Figures 1.6, 1.7, and 1.8
present the main features of his system; a colour plate detailing Munsell's hue, value,
and chroma sequences may be found in reference [13] (see Figure 1.5 therein).

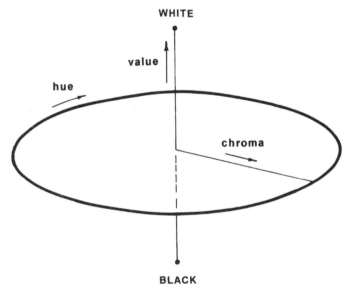

Figure 1.6   The Munsell colour system [11].

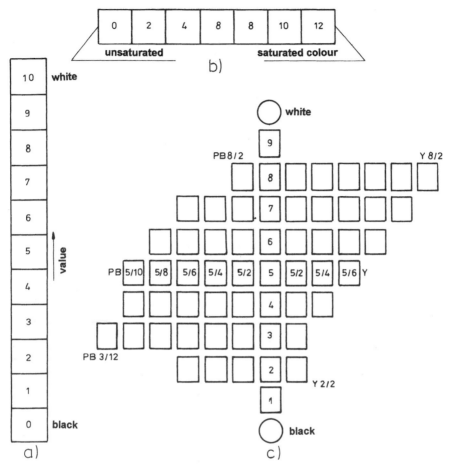

Figure 1.7 Basic elements of the Munsell system (omitting the hue circle) [11].

From the first Figure, it can be seen that the system is characterised by the hue circle and a perpendicular axis presenting the Munsell values. This achromatic axis comprises the values of perceptually equidistant grey contents, which are assigned integral numbers, from zero for pure black to 10 for pure white. In other words, **value** is an attribute corresponding to the lightness of a specific grey sample. **Chroma** is described by numbers 0, 2, 4,...,12, as shown in Figure 1.7b, from the centre towards the circle. It should be stressed that the maximum Munsell chroma for various hues is not equidistant in relation to the value axis. Hence, the Munsell colour space is irregular as seen in Figure 1.7c.

Figure 1.8 presents the Munsell hue circle. It was arbitrarily divided into ten geometrical equal sectors based on five 'principal hues': red, yellow, green, blue, and purple, i.e. four chromatic colours and one achromatic colour. The remaining five 'intermediate hues' are designated as yellow red, green yellow, blue green, purple blue, and red purple. All the hues were assigned symbols R, YR, Y, GY, G, BG, B,

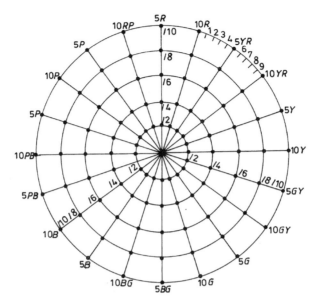

Figure 1.8 Arrangement of the hue circle in the Munsell system [6]. {Colours, clockwise from the top: R = red, Y = yellow, G = green, B = blue, P = purple}

PB, P, and RP. As can be seen from the Figure, a further subdivision can be achieved, e.g. 2.5YR is to be understood as a perceived colour between 2YR and 3YR. The Munsell system and the Munsell handbook of colour are referred to in many monographs and articles on colour science, but a full treatment of them lies beyond the scope of the present book.

1.5 QUANTITATIVE COLOUR EVALUATION – TRICHROMATIC COLORIMETRY

1.5.1 Additive colour mixing. CIE chromaticity diagram and coordinates

The quantitative measurement of colour is based on trichromatic colorimetry and Grassman's laws of additive colour mixing. According to [5] and [11], the three empirical laws which describe the properties of colour matching through the additive mixing of colour stimuli can be formulated as follows:

1. To reach a colour match, three independent variables are necessary and sufficient.

2. In additive mixtures of two colour stimuli, if one component is gradually changed, the resulting colour also changes gradually. This law expresses the fact that every colour has another neighbouring colour (as in the colour wheel).

3. The colour of an additive colour mixture depends only on the colour of its components, but not on their spectral compositions. Hence, only the tristimulus values (see below) of the components are relevant to this colour.

These basic laws make it possible to formulate mathematical relationships and to express them in an analytical form.

All these problems are described in many books and articles [2, 3, 4, 6, 11, 13, 14], with a recent review of CIE tristimulus colorimetry [15] providing copious references. Similarly detailed consideration is beyond the scope of the present book – only the most important features are tackled here.

To obtain an analytical expression of colour, it should be assumed that:

1. the identity of colour sensations resulting from colour mixing may be expressed as algebraic equality,

2. colour mixing itself may be expressed as a sum,

3. various proportions of colours involved in mixing may be expressed by numerical coefficients.

From Grassman's laws, it follows that only four linearly dependent colours determine the colour of the mixture. Thus

$$f'F = r'R + g'G + b'B. \tag{1.1}$$

R, G, and B denote three independent primary colours (a so-called colour triad). These can in fact be any colours such that none of them can be obtained by mixing the other two. The symbols r', g', b', and f' are the numerical coefficients for the colours identical with the colour of the components and of the mixture. These coefficients are also termed **colour modules**. Now, if the colours R, G, and B are assumed to be basic and unitary colours, the equation changes to:

$$F = rR + gG + bB, \tag{1.2}$$

where $r + g + b = 1$.

On the basis of Grassman's laws, a vector model of colour stimuli can be applied, as the three independent colours can be represented as vectors in a three-dimensional space with their beginnings at one point (which corresponds to the blackness point). As the representation of colour in space is rather complicated, a planar diagram may be constructed instead. For this purpose, an assumption is made that one of the colour attributes, lightness, could be represented by a point and the other two, corresponding to changes of hue and saturation, form the so-called **colour triangle**, which encompasses all colours characterised by different chromaticities.

The colour triangle is shown in Figure 1.9 [11]. The vertices of the triangle represent the chromaticities of the three colours, whereas the points on its sides, a, b, etc., correspond to mixtures of two of the colours. The nearer a given point is to a vertex, the greater is the proportion of the colour represented by that vertex. In practice, the three independent colours chosen in trichromatic colorimetry have the following wavelengths: R (red) = 700 nm, G (green) = 546.1 nm, and B (blue) = 435.8 nm.

To get a mixture of three colours, one proceeds step by step, as indicated in Figure 1.9. For instance, point a is obtained by mixing R and G, and then adding the colour

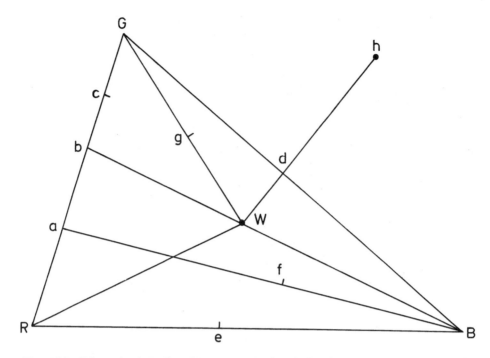

Figure 1.9   Colour triangle R, G, and B represent the three basic colours; W is the white point [11].

B to the mixture yields point *f* as the final result. If appropriate proportions of each of the colours are used, one can obtain white colour denoted by *W* in the figure. Further, it is clear that the *WR*, *WG*, and *WB* segments represent a gradual increase of colour saturation towards the corresponding vertex reaching the unitary value (in accordance with the assumption). However, the saturation can be greater than unity and such colours are outside the triangle plane. The quantities *r*, *g*, and *b* are given by the formulae:

$$r = \frac{R}{R+G+B} \quad g = \frac{G}{R+G+B} \quad b = \frac{B}{R+G+B}, \quad (1.3)$$

and are called trichromatic coordinates. They may be positive or negative, the first equal or less than 1, and according to (1.2) their sum must be equal to 1.

The tristimulus values *R*, *G*, and *B* can be expressed by the following formulae:

$$R = \int \varphi(\lambda)\bar{r}(\lambda)\mathrm{d}\lambda \quad G = \int \varphi(\lambda)\bar{g}(\lambda)\mathrm{d}\lambda \quad B = \int \varphi(\lambda)\bar{b}(\lambda)\mathrm{d}\lambda \quad (1.4)$$

for integration limits 380–780 nm,[4]

---

[4]   Recently it has been recommended that integration should be carried out up to 830 nm.

where $\bar{r}$, $\bar{g}$, and $\bar{b}$ are the so-called normalised colour-matching functions and $\varphi(\lambda)$ is the relative spectral distribution of the colour stimulus,

Colour matching functions are defined as the tristimulus values, with respect to the given three primary colours, of monochromatic lights of equal radiant energy, considered as functions of the wavelength.

In matching colours according to equation (1.1), it may sometimes happen that mixing R, G, and B colours does not yield a satisfactory result, i.e. the obtained colour is not exactly the same as the standard. Practically, this means that the addition of colours is not the appropriate method in such cases. To obtain the required result, one of the primaries should be mixed with the standard colour. This should be interpreted to mean that the given colour (situated outside the RGB triangle) can be matched by mixing a negative amount of one of the primary colours with positive amounts of the other two.

To avoid this situation, a transformation is made from the trichromatic system based on the primaries R, G, and B to a system based on new primaries X, Y, Z which cannot be realised by an actual light source [4]. These primaries are nonreal or imaginary. However, the triangle formed by the chromaticity points of X, Y, and Z contains the spectrum locus and the purple line. Hence, the new tristimulus values X, Y, and Z and the corresponding x, y, z of any real colour are always positive.

$\bar{x}$, $\bar{y}$, and $\bar{z}$ as functions of the wavelength are shown in Figure 1.10 and the numerical data are to be found in numerous monographs (according to CIE, for the standard observer, for a 2° or 10° angle of observation).

By analogy to (1.4), in the XYZ system we have:

$$X = k \int \varphi(\lambda)\bar{x}(\lambda)d\lambda \quad Y = k \int \varphi(\lambda)\bar{y}(\lambda)d\lambda \quad Z = k \int \varphi(\lambda)\bar{z}(\lambda)d\lambda, \quad (1.5)$$

where $\bar{x}$, $\bar{y}$, and $\bar{z}$ have already been defined and k is a normalizing factor given by the formula:

$$k = \frac{100}{\int S(\lambda)\bar{y}(\lambda)d\lambda}. \quad (1.6)$$

Now, considering the colour of an object which transmits or reflects light, it can be found that the spectral distribution of the colour stimulus is given by the product of the spectral transmission or reflection coefficients, $\tau(\lambda)$ or $\rho(\lambda)$, and the relative energy spectral distribution of the light source (illuminant), S. Accordingly, we obtain the following formulae:

for transmission $\quad \varphi(\lambda) = \tau(\lambda)S(\lambda);$ (1.7)

for reflection $\quad \varphi(\lambda) = \rho(\lambda)S(\lambda).$ (1.8)

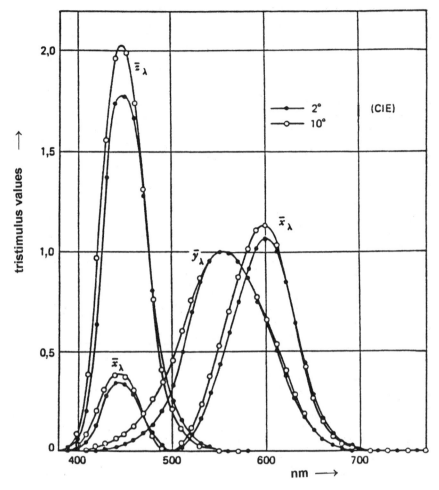

Figure 1.10 The CIE colour-matching function for the Standard Colorimetric Observer (2°: —•—; 10°: —o—).

In practice, summation is done instead of integration in accordance with the following:

$$X = k \sum_{\lambda_i} \varphi(\lambda_i)\bar{x}(\lambda_i)\Delta\lambda$$

$$Y = k \sum_{\lambda_i} \varphi(\lambda_i)\bar{y}(\lambda_i)\Delta\lambda \qquad (1.9)$$

$$Z = k \sum_{\lambda_i} \varphi(\lambda_i)\bar{z}(\lambda_i)\Delta\lambda,$$

where

$$k = \frac{100}{\sum_{\lambda} \varphi(\lambda_i)\bar{y}(\lambda_i)\Delta\lambda}. \qquad (1.10)$$

The intervals are usually 1.5 or 10 nm; in the experiments described in this book, a 1 nm interval was always applied.

As in the *RGB* system, these tristimulus values lead to **chromaticity coordinates** x, y, z:

$$x = \frac{X}{X+Y+Z} \quad y = \frac{Y}{X+Y+Z} \quad z = \frac{Z}{X+Y+Z}, \tag{1.11}$$

and, of course, their sum is $x + y + z = 1$. Hence, to characterise a colour, only two coordinates are needed, usually x and y. These coordinates are the basis of the chromaticity diagram shown in Figure 1.11 and in colour as Figure C1.

This is the well-known 'horse-shoe'-shaped **spectrum locus**, encompassing 23 colour fields, which are given proper names as used in the English language. The spectrum locus comprises all chromatic colours. As no colours can be of greater saturation than the chromatic ones, it is an envelope of all physically real colours. If one considers the colour space instead of the plane diagram, all colours will be

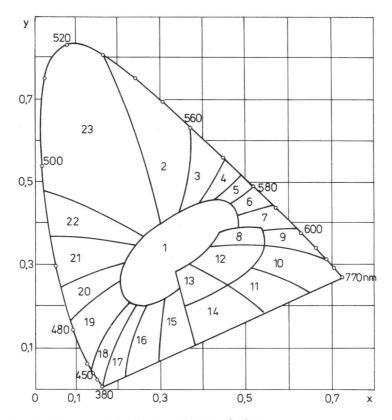

Figure 1.11 CIE 1931 chromaticity diagram and names of colours:
1 – white; 2 – yellowish green; 3 – yellow green; 4 – greenish yellow; 5 – yellow; 6 – yellowish orange; 7 – orange; 8 – orange pink; 9 – reddish pink; 10 – red; 11 – purplish red; 12 – pink; 13 – purplish pink; 14 – red purple; 15 – reddish purple; 16 – purple; 17 – bluish purple; 18 – purplish blue; 19 – blue; 20 – greenish blue; 21 – blue green; 22 – bluish green; 23 – green. From p. 50 of ref. [4]; a coloured version of this diagram appears as Figure C1 in the centre of this book.

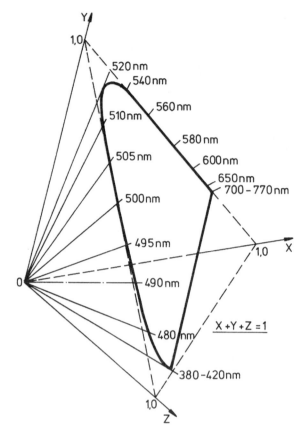

Figure 1.12 Colour solid and plane section, with corresponding monochromatic wavelengths [2].

contained in a cone whose vertex is the origin of the system and the generating lines correspond to the chromatic colours. The colour space is shown in Figure 1.12.

The straight line between 380 nm and 770 nm is called the **purple line** and the triangle formed by these points and the point corresponding to the coordinates of the illuminant used is termed the **purple triangle** (Figure 1.13). It should be remembered that purple colours are achromatic.

The x, y chromaticity coordinates are not sufficient to express all attributes of colour and, as has been stated, they are to be correlated with hue and saturation. The third value, lightness, is given by the tristimulus value Y, which in the CIE 1931 system is the measure of luminance or, more exactly, of the luminance factor in relation to a white standard. The Y value can be visualised on an axis perpendicular to the chromaticity plane. A diagram calculated on the basis of simulated spectra and x, y, Y values is shown in Figure 3.23 in Chapter 3.

It is important to realise that the chromaticity diagram represents colour stimuli and not perceived colours. As the aim of this book is to show the usefulness of chromaticity coordinates to create a new source of information about chemical

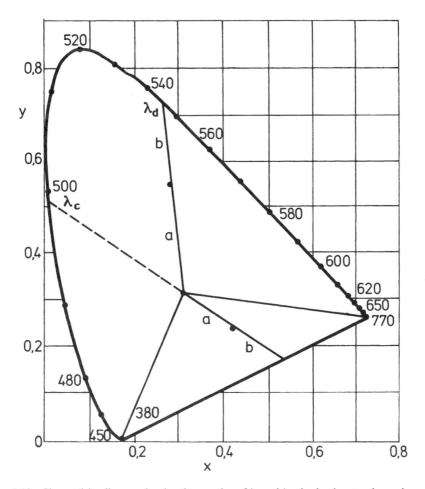

Figure 1.13 Chromaticity diagram showing the meaning of λ_d and λ_c, the dominant and complementary wavelengths (in nm).

compounds, colour stimuli and not perceived colours are of primary importance. The latter are strongly dependent on the observer and different viewing conditions. Hence, the x, y coordinates determine one and only one point on the diagram, under strictly defined measuring conditions, and they should be treated as characteristic of particular coloured objects. Besides, additional information can be derived from the chromaticity diagram.

As can be seen in Figure 1.13, the straight line connecting the point which corresponds to the illuminant (assumed to be an achromatic colour stimulus) with the point representing the given coordinates (colour) intersects the spectrum locus at a particular point, which determines the **dominant wavelength**. If, however, the colour of an object is situated within the purple triangle, such an intersection defines the **complementary wavelength**. The ratio $a/(a + b)$ is called **excitation purity** (which has been defined above). Using the luminances of the specimen and of the white standard, one can also define **colorimetric purity** as:

$$p_c = \frac{L_d}{L_d + L_n}. \tag{1.12}$$

The interrelation between p_e and p_c is given by the formula

$$p_c = p_e y_d / y, \tag{1.13}$$

where y_d and y are the y coordinates of, respectively, the chromatic stimulus and the colour stimulus.

The CIE chromaticity diagram suffers from an important drawback; namely, the distribution of colours is non-uniform. This feature can be seen if one considers the differences between colours. Such differences are shown as short lines joining two colours. Assuming the same luminances of all colours and the same magnitude of perceptual colour differences, all lines in Figure 1.14 should be of the same length. They are, however, much longer in the green part of the diagram and much shorter near the violet end than the mean length.

Formally, it is not possible to avoid this problem completely, as it is not possible to represent a curved surface on a plane. A more exact representation, from this point of view, has been introduced as the Uniform Chromaticity Diagrams, and CIELAB and CIELUV spaces.

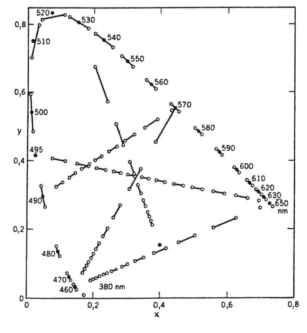

Figure 1.14 CIE chromaticity diagram showing lines which correspond to non-uniform distribution of colours (see text for explanation).

1.5.2 Improved CIE systems: CIELAB and CIELUV

One of the standard diagrams is the CIE 1976 uniform chromaticity scale (UCS) diagram also referred to as the u', v' diagram [5]. The new coordinates are formulated as follows:

$$u' = \frac{4X}{X + 15Y + 3Z} = \frac{4x}{-2x + 12y + 3};$$ (1.14)

$$v' = \frac{9X}{X + 15Y + 3Z} = \frac{9x}{-2x + 12y + 3}.$$ (1.15)

It can be shown that the differences between the lengths of the lines shown in Figure 1.14 are considerably smaller and hence the diagram is said to be more uniform.

Even more important to the practical use of colorimetry are the **uniform colour spaces**, which – in contrast to the CIE diagram – reflect the luminances of coloured objects in addition to their chromaticities. The use of these three dimensions is particularly valuable for the evaluation of colour differences.

There are two such spaces, CIELAB and CIELUV. The respective coordinates are given by formulae (1.16) to (1.21):

a) the CIELAB or CIE 1976 $L^*a^*b^*$ system,

$$L^* = 116(Y/Y_n)^{1/3} - 16;$$ (1.16)

$$a^* = 500\left[(X/X_n)^{1/3} - (Y/Y_n)^{1/3}\right];$$ (1.17)

$$b^* = 200\left[(Y/Y_n)^{1/3} - (Z/Z_n)^{1/3}\right],$$ (1.18)

where X_n, Y_n, and Z_n denote trichromatic values for the white standard.

b) the CIELUV or CIE 1976 $L^*u^*v^*$ system,

$$L^* = 116(Y/Y_n)^{1/3} - 16;$$ (1.19)

$$u^* = 13L^*(u' - u'_n);$$ (1.20)

$$v^* = 13L^*(v' - v'_n),$$ (1.21)

where u'_n and v'_n are the u' and v' values for the white standard according to formulae (1.14) and (1.15).

Differences between two colours in these spaces are expressed by (1.22) and (1.23):

a) in the CIELAB space,

$$\Delta E^*_{ab} = \left[(\Delta L^*)^2 + (\Delta a^*)^2 + (\Delta b^*)^2\right]^{1/2};$$ (1.22)

b) in the CIELUV space,

$$\Delta E^*_{uv} = \left[(\Delta L^*)^2 + (\Delta u^*)^2 + (\Delta v^*)^2\right]^{1/2}.$$ (1.23)

For a better characterisation of colour, some other quantities are also used. These are the so-called **hue angle**, h_{ab} and h_{uv}:

$$CIE\ 1976 - h_{ab} \qquad h_{ab} = \text{arc } \tan(b^*/a^*); \tag{1.24}$$

$$CIE\ 1976 - h_{uv} \qquad h_{uv} = \text{arc } \tan(v^*/u^*), \tag{1.25}$$

and saturation:

$$S_{uv} = 13\left[(u' - u'_n)^2 + (v' - v'_n)^2\right]^{1/2} \tag{1.26}$$

As stated, the value of the hue angle corresponds more exactly to the perceived colour than the dominant wavelength determined by the x, y chromaticity coordinates. It has also been calculated for some systems discussed in this book. The h_{ab} hue angle determined in the course of investigating the solvent effect for Co(II) solutions allowed us to distinguish clearly between blue and pink solutions. As is well known, the former contain mainly tetrahedral Co(II) forms, while the latter contain almost exclusively or at least predominantly octahedral complex ions (Chapter 5).

1.6   METAMERISM

Although the absorption and reflectance spectrum is the basis for calculating the chromaticity coordinates of an object, hence determining its colour (or at least the colour field on the chromaticity diagram), the situation is not reversible because the full envelope of the spectrum cannot be predicted from the set of coordinates. The direct result of this situation is the so-called **metamerism**: even when two objects are seen as being of the same colour, i.e. their tristimulus values are identical under one source, they are said to be metameric if their absorption spectra differ from each other.

The identity of colour may be simply expressed as follows:

$$\begin{aligned} X_1 &= X_2 \\ Y_1 &= Y_2 \\ Z_1 &= Z_2. \end{aligned} \tag{1.27}$$

Values of X, Y, and Z are formulated by appropriate integrals or, in actual practice, by corresponding sums (cf. Section 1.9) which contain, among other parameters, the characteristic spectral light distribution of the light source used. If another source is used instead, the colours will no longer be matched.

Changing the light source (illuminant) is in practice used when considering the problem of the metamerism of two objects. This phenomenon is of great importance both in industry and in everyday life, examples including familiar situations such as using a paint in day and night light, or of choosing or matching coloured fabrics in daylight or artificial light.

Table 1.3 presents examples of three metamers and their chromaticity coordinates and luminances for two different illuminants, D_{65} and A [9].

Table 1.3 Chromaticity coordinates of 3 metamers for illuminants D_{65} and A [3]

		D_{65}	A
	x_1	0.4691	0.5680
1	y_1	0.3643	0.3847
	Y_1	33.0	40.25
	x_2	0.4691	0.5683
2	y_2	0.3643	0.3810
	Y_2	33.0	40.23
	x_3	0.4691	0.5592
3	y_3	0.3643	0.3941
	Y_3	33.0	40.36
$\Delta E(1 \leftrightarrow 2)$		0.0	2.7
$\Delta E(1 \leftrightarrow 3)$		0.0	11.2

It is generally concluded that the most frequently encountered metamerism is due either to the illuminant or to the observer, and if they are changed, metamerism disappears. In some cases, the situation becomes more complicated and even a change of the light source does not cancel this phenomenon. This happens when the spectra of both objects cross at at least three points or, more exactly, if the colour stimuli show this feature.

This kind of pair of objects is referred to as **metameric objects** or a **metameric pair**. Formally, one has to imagine a situation when even for three or more illuminants, the pair is still metameric. Accordingly, the number of colour stimuli crossing points must be greater. In other words, the stronger the similarity between two spectra (i.e. the same ordinate values at the same wavelengths), the greater the number of illuminants for which the objects are still metameric. In the limiting case, if two objects match invariably under all sources, they must be characterised by the same spectra.

In order to differentiate between various metameric pairs, it is useful to formulate a measure of the difference of the spectra. Such a measure is termed the **index of metamerism**. Two main indices are known: the CIE **Illuminant Metamerism Index**, M, and the so-called **Observer Metamerism Index** [6]. In the former, the colour difference between a metameric pair is achieved by changing the test illuminant to a reference illuminant (of different spectral composition). Two sets of tristimulus values X_1, Y_1, Z_1 and X_2, Y_2, Z_2 are used to calculate the colour difference (and hence the metameric index) according to the CIE difference formulae.

A general formula for the metamerism measure was given by Bartelson [1]:

$$D = \left\{ \sum_{\lambda} [\varphi_1(\lambda) - \varphi_2(\lambda)]^2 \right\}^{1/2}, \qquad (1.28)$$

where φ_1, φ_2 denote the colour stimuli functions of the metamers.

An interesting example of metamerism is connected with the colour pink. Pink light may be obtained in various different ways by mixing appropriate light beams

[16] – additive mixing of red and white, of red and cyan, or of red and green and violet yields metameric beams of pink light.

Now, when one considers the pink colour of an object, the same situation occurs. Hence, it is impossible to predict not only the spectrum envelope but even the number of bands. An important practical case occurs in the chemistry of transition metals. Octahedral Co(II) compounds are pink in aqueous solutions as well as in many organic solvents showing only one absorption band at ca. 500 nm and strong transmission in other regions of the spectrum. The problem is discussed in more detail elsewhere in the book (mainly in Chapter 3). But a pink colour is also characteristic of, for instance, a chromate salt in a white matrix, although the electronic spectrum of the chromate ion is completely different from that of a Co(II) octahedral ion.

It should be stressed that metamerism is a formal obstacle to the use of chromaticity coordinates in describing the properties of an object (chemical compound) under study. It is clear that the set of three numbers defining colour – x, y, Y or X, Y, Z or dominant wavelength and purity (saturation) – contain less information than a complete absorption or reflectance spectrum. The latter constitutes a primary source of information; chromaticity coordinates are a secondary one.

However, the significance of quantitative colour characteristics increases considerably when they are used for a limited class of compounds, e.g. the chemical compounds of the same element. In this book, transition metals are the principal object of interest. Spectroscopic properties in the visible range clearly demonstrate that characteristic colour is chiefly due to a part of the compound called the chromophoric group. Hence, its colour given by chromaticity coordinates can serve as an additional source of information. The problem is discussed further in the remainder of the book.

1.7 FURTHER READING

We have in this short introductory chapter presented a number of fundamental matters and issues rather briefly. Readers wishing for more background and context, presented at a similar level, are advised to consult articles on colour in scientific and technological encyclopaedias – references [13, 17, 18] are particularly recommendable. For fuller treatments the reader should, of course, consult the major texts devoted to the chemistry, physics, and technology of colour and colour perception [3, 4, 6, 11, 14, 16, 19, 20], several of which have already been cited at appropriate points. For details of illuminants, detectors, CIE and colour order systems, and for formulae and data treatment, the reader may be directed to Wyszecki and Stiles's technical text [21]. Koenderink [8] lists a number of colour atlases, which may be supplemented by the Colour Index [22][5] and other publications by the Society of

---

[5] The *Methuen Handbook of Colour*, published by Eyre Methuen (e.g. 3rd edn., 1978), contains not only an extensive atlas of colour samples and names, but also connects these to the colour circle, to the CIE and Munsell systems, to the British Standards Institution's colour specifications, and to names employed by paint manufacturers.

Dyers and Colourists. Stiles [23] and Boynton [24] deal with human colour vision, while Hunt's book on colour reproduction [25] is obviously of particular relevance to the communication of colour information through the medium of the printed page. As commented in passing in the final Chapter of this book, the value of coloured illustrations in enhancing the presentation of, for example, transition metal chemistry in textbooks may be considerably negated by poor colour reproduction.

REFERENCES

[1] Bartelson, C.J., 1980, *Colorimetry*, in *Optical Radiation Measurements*, Vol. 2, Academic Press, New York.

[2] Felhorski, W. and Stanioch, W., 1973, *Kolorymetria trójchromatyczna*, WNT, Warsaw.

[3] Judd, D.B. and Wyszecki, G., 1975, *Color in Business, Science and Industry*, 3rd edn., Wiley, New York.

[4] Billmeyer, F.W. and Saltzman, M., 1981, *Principles of Color Technology*, 2nd edn., Wiley, New York.

[5] *International Dictionary of Lighting Techniques, 1970, Publ. No. 17-E-1.1*: International Lighting Vocabulary, 3rd edn.

[6] Hunt, R.W.G., 1987, *Measuring Colour*, Ellis Horwood, Chichester; 1992, 2nd edn, Wiley, New York.

[7] McCamy, C.S., 1985, *Color Res. & Appl. (USA)*, **10**, 20.

[8] Koenderink, J.J., 1987, *J. Opt. Soc. Am. A*, **4**, 1314.

[9] Dordet, Y., 1990, *Colorimetrie. Principles et Applications*, Eyrolles, Paris.

[10] Munsell, A.H., 1990, *A Color Notation*, 14th edn., Macbeth Division of Kollmorgen Instruments Corporation, Newburgh, New York; *Munsell Book of Color*, continuously available from the Munsell Color Co., Baltimore, Md., from 1929, then from Macbeth Division of Kollmorgen Instruments Corporation, Newburgh, New York, 1946–1961.

[11] Zausznica, A., 1959, *Nauka o barwie*, PWN, Warsaw.

[12] Jacobson, E., 1948, *Basic Color: An Interpretation of the Ostwald Color System*, Theobald, P., Chicago, Ill.

[13] Nassau, K., 1996, *Color*, in *Kirk-Othmer's Encyclopaedia of Chemical Technology*, 4th edn., Wiley, New York, Vol. 6, p. 841.

[14] Richter, M., 1976, *Einführung in der Farbmetrik*, Walter de Gruyter, Perelin.

[15] Krishna Prasad, K.M.M., Raheem, S., Vijayalekshmi, P. and Kamala Sastri, C., 1996, *Talanta*, **43**, 1187.

[16] Nassau, K., 1983, *The Physics and Chemistry of Color*, Wiley, New York.

[17] Billmeyer, F.W., 1979, *Color*, in *Kirk-Othmer's Encyclopaedia of Chemical Technology*, 3rd edn., Wiley, New York, Vol. 6, p. 523.

[18] Boynton, R.M., 1997, *Color*, in *McGraw-Hill Encyclopaedia of Science & Technology*, 8th edn., McGraw-Hill, New York, Vol. 4, p. 178.

[19] Wright, W.D., 1969, *The Measurement of Colour*, 4th edn., Adam Hilger, London.

[20] McLaren, K., 1986, *The Colour Science of Dyes and Pigments*, 2nd edn., Adam Hilger, Bristol.
[21] Wyszecki, G. and Stiles, W.S., 1967, *Color Science – Concepts and Methods, Quantitative Data and Formulas*, Wiley, New York.
[22] *The Colour Index*, 1971, 3rd edn., Society of Dyers and Colourists, Bradford/ London.
[23] Stiles, W.S., 1978, *Mechanisms of Color Vision*, Academic Press, New York.
[24] Boynton, R.M., 1979, *Human Color Vision*, Holt, Reinhart and Winston, New York.
[25] Hunt, R.W.G., 1975, *The Reproduction of Colour*, 3rd edn., Fountain Press, London.

2. METAL COMPOUNDS AS COLOURED OBJECTS

As already established in Chapter 1, the colour stimulus function is the product of three parameters. These are the spectral distribution function of the light source (the illuminant), the object's reflection or absorption spectrum, and the spectral distribution function of the sensitivity of the eye.

Concentrating on the role of the object, we may identify some basic physical causes and mechanisms of colour production in such coloured objects. Nassau [1] distinguished fifteen causes of colour formation connected with interactions between light and molecules, atoms, and electrons. These include ligand field transitions, transitions between molecular orbitals, interference, and light diffusion. The chapters of Nassau's book which are most relevant to our present book are those on colours arising from ligand field transitions (his Chapter 5), from charge-transfer transitions (his Chapter 7), and from colour centres (his Chapter 9). Table 2.1 lists, with examples, the principal sources of colour in metal compounds discussed in our

Table 2.1 Various causes of colour in metal compounds, with examples

Transitions	Examples
d-d: crystal (ligand) field	transition metal compounds, e.g. $CoCl_2$
	some minerals, e.g. Cr_2O_3, $MnCO_3$
	some mineral pigments, e.g. Fe_2O_3
	some minerals and gemstones with transition metal ions as impurities, e.g. ruby ($Al_2O_3 + Cr^{3+}$)
f-f	lanthanide compounds, e.g. $NdCl_3$
	actinide compounds, e.g. UCl_4
	some minerals with lanthanide impurities
Charge-transfer[a]	metal oxoanions, e.g. MnO_4^-
	some minerals, e.g. $PbCrO_4$
	d- and *f*-block metal complexes with organic ligands such as acetylacetonate
Between energy bands	colour centres, e.g. in alkali metals
	many metal sulphides, e.g. PbS

[a] These may be metal-to-ligand charge-transfer, MLCT, ligand-to-metal charge-transfer, LMCT, or metal-to-metal intervalence charge-transfer, MMCT or IVCT.

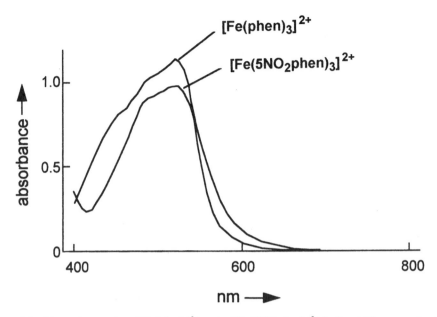

Figure 2.1   Absorption spectra of [Fe(phen)₃]²⁺ and of [Fe(5NO₂phen)₃]²⁺ in the visible range, showing identical λ_{max} values but different overall contours – the significantly different colours of solutions containing these complexes are shown in Figure C2.

present book. Our main focus is on ligand field (crystal field) transitions, which occur not only in pure transition metal compounds but also in minerals, pigments, and glasses doped with such compounds, and on charge-transfer transitions.

It is possible to determine the colour of a chemical compound – apart from a purely subjective evaluation – by means of collections (atlases) of colour samples. However a central theme of this book is to discuss the problems inherent in quantitative colour measurement, based on recorded electronic spectra, in particular on absorption (transmission) spectra. It is important to record and evaluate spectra over the complete visible range, and not just to rely on wavelengths (wavenumbers) of maximum (minimum) absorption. This is well illustrated by two iron complexes, [Fe(phen)₃]²⁺ and [Fe(5NO₂phen)₃]²⁺, whose colours differ significantly (Figure C2) despite identical maximum absorption wavelengths (λ_{max}). Figure 2.1 shows their identical λ_{max} values but different spectral contours in the visible region.

2.1   GENERAL ISSUES

These issues are important when one tries to determine and specify precisely the colour of an object under study on the basis of the interpretation of absorption or reflection spectra given in the literature. Even authors of publications reporting new chemical compounds generally pay scant attention to this matter.

As is well known, absorption bands can be characterised by the following para-meters: band position, half-width, and intensity expressed as the molar absorption coefficient (or more precisely as oscillator strength – which also comprises half-width), as well as band shape. It should be noted that reflection spectra are usually significantly different in shape from the respective absorption spectra.

Let us consider the effect of the individual features on the colour of an object:

1. The form of the band does not directly determine colour or give grounds for a firm prediction of colour, yet it does have some informative value. First of all, one may distinguish between perfectly (or almost perfectly) symmetrical bands and asymmetrical ones. In the case of asymmetrical bands in the borderline region between the visible range and ultraviolet, the long-wave arm can reach so far into the visible range that the corresponding light absorption significantly affects the colour of the object. Additionally, in transition metal compounds, charge-transfer (CT) transitions often occur in the range under consideration. CT transitions are characterised by large molar extinction coefficients, of the order of 10^3 to 10^4 as compared with values of 10^1 to $10^2 \, M^{-1} \cdot cm^{-1}$ for crystal field transitions. A small concentration of such a coloured object, as doping agent or impurity, may thus play a decisive role in determining colour. CT transitions generally give strong absorption in the blue region of the spectrum.

An asymmetrical band may also occur in the borderline region between the visible range and infrared. However, in the absorption spectra of transition metal compounds such bands are usually weak (as crystal field, CF, bands) and asymmetry is of little consequence here. But in the case of glass doped with transition elements these bands are considered to have a certain significance (Chapter 8).

It should be stressed that although colour is connected with the envelope, an interpretation aiming at relating colour to specific electronic transitions requires a computer-aided resolution into component bands, particularly in the case of asymmetrical bands. Failure to do this often precludes a reliable interpretation, unless one also has other physicochemical data (e.g. those obtained from the examinations of single crystals at low temperatures).

2. Of course, the position of an absorption band defines directly the colour of an object as a complementary colour if it is the only one in the visible range. This is shown schematically in Figure 2.2 [2]. It should, however, be recalled that the spectra of transition metals usually show more than one absorption band and then, in principle, it is not possible to predict a specific colour. Some tentative conclusions can be drawn on the basis of the position of the absorption minimum. If there are several minima, the colour may be predicted approximately by considering additive colour mixing. Some important issues connected with the position of absorption bands stem from the theory of transition metal electronic spectra, which is discussed in Chapter 3. The characteristic quantity in that theory is the splitting parameter, Dq.

The simulations described in Chapter 3 show that the Dq parameter may to a certain extent be used to predict the colour of a compound, and its shifts if it is considered for a series of changing values. As a first approximation, this parameter is a function of spin-allowed transition energies, the number of which transitions depends on the crystal field symmetry and increases as the symmetry decreases. The situation can be presented most easily for octahedral complexes using the spectrochemical series. Increasing the Dq value causes a hypsochromic shift and decreasing Dq brings about a bathochromic shift of the entire spectrum, but mostly of the spin-allowed transition maxima, as some spin-forbidden transitions do not depend on Dq.

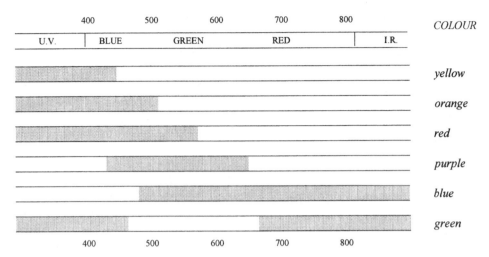

Figure 2.2   Colour appearance as a result of absorption of spectral radiation in the range 300 to 900 nm; denotes absorption, _____ denotes transmission [2].

This can be seen from the Tanabe-Sugano diagrams, where dependencies corresponding to the latter are straight lines [3]. In the case of small or large Dq values, the spectrum may be shifted so far towards the long-wave or the short-wave end that some absorption bands occur in a range that has little effect on the colour of an object.

However, the most important aspect is the fact that the colour of an object may change cyclically, that is in accordance with the colour wheel. This issue is discussed in more detail in Chapter 3 in connection with the results of the determination of the colour of Cr(III) compounds as a function of the Dq value. It can be concluded that if changes of Dq do not lead to a marked bathochromic or hypsochromic shift of the bands and the consequent change of the number of absorption bands, the cyclicity of colour shifts should be preserved.

It follows that non-continuous changes can be linked with a continuous colour change as the latter may at best be cyclic. Thus, for instance, in studies of solvatochromism (Chapters 3 and 5), it is often shown that the wavelength (or wavenumber) of a given transition is correlated with the acceptor or donor number of the solvent and the occurrence of a specific colour. However, the colour of a solution of a compound cannot be predicted by extrapolation of this dependency.

The significance of the Dq parameter is not so clear in the case of the colour of compounds in the solid phase. This issue is described in more detail in Section 3.3.2.3; here, we should recall that the Dq value also depends on the central ion-ligand distance, and in the solid phase on structural factors as well as on deviations from the ideal stoichiometry of the compound. This has been shown, for instance, for NiO and various oxide systems containing Cr^{3+} (Chapter 3).

3. The intensity of absorption bands, expressed as either the molar absorption coefficient of the individual absorption bands or the oscillator strength of those bands, is important for the interpretation of structural aspects concerning the

geometry of a complex, symmetry, and electronic configuration (high-spin or low-spin). Perceived colour depends on the concentration of, for instance, a compound in solution or the percentage of chromogenic factor in a matrix, such as glass or a mineral pigment. Thus absorbance changes may cause a detectable colour change, even though the overall form of the spectrum does not change. Another complicating factor here is so-called dichromatism [4], as a result of which a change in the concentration of an object may change its colour (in the chromaticity diagram – colour field), sometimes quite dramatically.

4. Half-width is an important characteristic of an absorption band. Its theoretical aspect is connected with the system of the potential curves (or more exactly potential surfaces) of the ground state and the excited state. The half-width of a given band corresponds to the slope of the excited state potential curve at the point corresponding to the minimum of the ground state curve. The significance of the width, as well as intensity, of absorption bands has been studied by means of computer simulation (see Section 3.3.1.1).

On the basis of simple premises, we can expect that the greater the half-width of a band (for the same wavelength of the maximum), the greater the total absorption of visible light and the smaller the luminance will be. According to Jørgensen [3], half-width may be roughly estimated as 0.3Δ or approximately $3Dq$ (for spin-allowed bands).

2.2 SOME MEASURING ISSUES

The most common form of representing electronic spectra in the visible range, particularly for solutions, is the absorption (or less often transmission) spectrum, that is in the absorbance-wavelength (nm) system. Chromaticity coordinates (formulated as in Chapter 1) for a given light source (illuminant) and colour system may be calculated by appropriate computer programmes. One author (A.B.) has used in his research the programme COLOR (Scheme 2.1) for absorption spectra and REFLEKS for reflectance spectra (of pigments and chemical compounds) [5, 6, 7, 8].

Measurements were done on a two-beam Hitachi spectrophotometer, model 356, coupled with an IBM/PC microcomputer.

The set of absorbance values is transposed to transmittance values, which is directly connected with the fundamental laws of light absorption and the Bouguer-Lambert-Beer law. It should be noted that too small an absorbance value is disadvantageous, especially in the CIE system, as the values obtained for the chromaticity coordinates are often located within the white (colourless) area in Figures 1.11 and C1. In this respect the CIELAB system, with $+a^*$, $-a^*$, $+b^*$, and $-b^*$ coordinates, is preferable.

A change in absorbance can be brought about by changing the concentration or the thickness of a layer, but – as already noted – this has a definite effect on the colour (for physical reasons): a change of concentration may also cause physicochemical changes in the solution (particularly if a complexing non-aqueous solvent is used).

The precision of measurements and the obtained chromaticity coordinates of the colour may be tested with two solutions recommended in the literature: filters which

Flow chart of the computer program COLOR
for the calculation of chromaticity coordinates

Read spectrum (1 or 2, 3, 4)
1. experimental absorption spectrum
2. experimental transmittance spectrum
3. experimental reflectance spectrum
4. simulated absorption spectrum

Print spectrum

Transformation of spectral data

Read
1. Spectral energy distribution of the light source
2. Colour-matching functions x, y, z

Selection of data
1. Range of spectrum
2. Intervals of wavelength

Calculate tristimulus values X, Y, Z

Calculate CIE chromaticity coordinates x, y, z

Calculate $CIELAB$ coordinates L^*, a^*, b^*

Calculate $CIELUV$ coordinates L^*, u^*, v^*

Print results

Scheme   2.1

in combination with a source A (a specific light-bulb) forms a standard light source
C. The following solutions are used for this purpose (a layer of 1 cm each):

Solution C_1	$CuSO_4 \cdot 5H_2O$	3.412 g
	$C_6H_8(OH)_6$ (mannitol)	3.412 g
	Pyridine	30 cm^3
	H_2O (dist.)	to 1 dm^3
Solution C_2	$CoSO_4 \cdot (NH_4)_2SO_4 \cdot 6H_2O$	30.58 g
	$CuSO_4 \cdot 5H_2O$	22.52 g
	H_2SO_4 (density 1.835 g.cm^{-1})	10 cm^3
	H_2O (dist.)	to 1 dm^3

D_{65} was used in measurements as the standard illuminant. Its spectral power distribution is shown in Figure 1.3. The following solutions corresponding to D_{65} [9] are used as chromaticity coordinate standards:

1. 0.04 g K_2CrO_4 in 1 dm^3 of 0.05 mol dm^{-3} KOH solution (chromaticity coordinates $x = 0.3160$, $y = 0.3280$, $Y = 0.9998$);

2. 20.0 g $CuSO_4 \cdot 5H_2O$ in 1 dm^3 water containing 10 cm^3 of concentrated H_2SO_4 (chromaticity coordinates $x = 0.2896$, $y = 0.3134$, $Y = 0.9294$);

3. 14.181 g $Co(NH_4)_2(SO_4)_2 \cdot 6H_2O$ in 1 dm^3 water containing 10 cm^3 of concentrated H_2SO_4 (chromaticity coordinates $x = 0.3314$, $y = 0.3112$, $Y = 0.8466$).

These three solutions were used in the author's (A.B.) own studies of colour, in which measurements were taken at 1 nm intervals.

If CIE chromaticity coordinates are plotted on the x, y chromaticity diagram, the dominant or complementary wavelength as well as excitation purity can be determined, as indicated in Figure 1.13. Although the CIE 1931 chromaticity diagram is conventionally divided into 23 colour fields, the set of x, y chromaticity coordinates is the basis for comparing the determined psychophysical colours of various coloured objects. A more precise measure is to use the difference ΔE^* based on CIELAB chromaticity coordinates (cf. Chapter 1).

The calculation of chromaticity coordinates for solids is more complicated and less precise. If reflectance spectra are measured, then first of all one has to take into account the lack of a linear dependency between the value of light reflection and the concentration of the substance, as well as the phenomena of light diffusion and absorption in the sample.

To calculate the chromaticity coordinates of generally opaque solids, the Kubelka-Munk theory [10, 11] is usually applied, which is also described in many monographs, e.g. [12, 13].

Assuming that when the sample (e.g. a pigment) is illuminated, light does not penetrate the deeper layers of the object, the Kubelka-Munk theory formulates the following dependency between the reflection, absorption, and diffusion coefficients:

$$\frac{(1 - R)^2}{2R} = \frac{K}{S},$$

where K denotes the absorption coefficient, S the diffusion coefficient, and R the reflection coefficient. The K/S value is characteristic of a particular substance at a given wavelength.

Duncan's equation [14] is also important:

$$\frac{K}{S} = \frac{c_1 K_1 + c_2 K_2 + \ldots + c_n K_n}{c_1 S_1 + c_2 S_2 + \ldots + c_n S_n}.$$

It describes mixtures of pigments or mixtures of a pigment and a matrix (i.e. a white substance, such as lithium carbonate, barium sulphate, or titanium dioxide). In this equation, c_i denotes the concentration of each of the ingredients, and K_i and S_i stand for, respectively, their absorption and diffusion coefficients. Various forms of the Kubelka-Munk equation are given, for example, in [15].

The interpretation of this simple equation leads to several important conclusions. If absorption, K, increases (e.g. due to increased amount of a pigment in admixture with a matrix), but the diffusion coefficient remains constant, then R decreases. Thus adding, for instance, a black pigment to a mixture reduces the reflection of light by the system. If, however, S increases while K remains unchanged, R increases too. A situation of this kind occurs if a strongly diffusive white pigment is added to a mixture. A thorough knowledge of absorption and diffusion spectra may serve as a basis for the selection of various pigments and matrices in mixtures to obtain the required value of reflection in a particular spectral range. Extensive examples are to be found in [15].

The chromaticity coordinates in the CIELAB system, a^* and b^*, represented as in Figure 5.8, are a practical indication of the particular proportions of four colours: blue on the $-b^*$ axis, yellow on the $+b^*$ axis, red on the $+a^*$ axis, and green on the $-a^*$ axis.[1] In practice, the yellow index ($Y.I.$) is used [11], which may be determined from the values of the trichromatic coordinates X, Y, Z, according to the following formula

$$Y.I. = \frac{128X - 106Z}{Y}.$$

In studies of the red colour of soil samples containing iron oxides [16] (mostly hematite and goethite), the authors used the red index, which may be determined from the CIELAB and CIE chromaticity coordinates. In the former system, the red index is calculated according to the following formula:

$$R_{Lab} = \frac{a^*(a^{*2} + b^{*2})^{1/2}}{b^* L^{*6}}.$$

It should be stressed that CIELAB a^* and b^* values have frequently been used in our studies on the colour of metal compounds – the next Chapter presents many examples. Apart from these parameters, applications of the hue angle, h_{ab}, and the colour difference, ΔE^*, have also been demonstrated – these are useful especially to characterise and differentiate colours of solutions.

---

[1] This is, as has been mentioned, the so-called tetrachromacy effect.

REFERENCES

[1] Nassau, K., 1983, *The Physics and Chemistry of Color*, Wiley, New York; Nassau, K., 1996, *Color*, in *Kirk-Othmer's Encyclopaedia of Chemical Technology*, 4th edn., Wiley, New York, Vol. 3, p. 841.

[2] Zausznica, A., 1959, *Nauka o barwie*, PWN, Warsaw.

[3] Jørgensen, C.K., 1962, *Absorption Spectra and Chemical Bonding in Complexes*, Pergamon, Oxford.

[4] Griffiths, J., 1976, *Colour and Constitution of Organic Molecules*, Academic Press, London.

[5] Bartecki, A., Tłaczała, T., Myrczek, J. and Raczko, M., 1990, *Barwne związki metali i ich zastosowanie* (*Coloured metal compounds and their applications*), Report No. 19/90, Politechnika Wrocławska, Wrocław.

[6] Bartecki, A. and Tłaczała, T., 1990, *Spectroscopy Lett.*, **23**, 727.

[7] Bartecki, A., Tłaczała, T. and Raczko, M., 1991, *Spectroscopy Lett.*, **24**, 559.

[8] Myrczek, J., unpublished computer program, 1988 (enquiries to the Institute of Inorganic Chemistry and Metallurgy of Rare Elements, Technical University of Wrocław).

[9] Kotrly, S. and Vytras, K., 1971, *Talanta*, **18**, 253.

[10] Kubelka, P. and Munk, F., 1931, *Z. Tech. Phys.*, **12**, 593.

[11] Kubelka, P., 1948, *J. Opt. Soc. Amer.*, **38**, 448.

[12] Billmeyer, F.W. and Saltzman, M., 1981, *Principles of Color Technology*, 2nd edn., Wiley, New York.

[13] Kortüm, G., 1969, *Reflexionspektroskopie*, Springer-Verlag, Berlin.

[14] Duncan, R.D., 1940, *Phys. Soc. Proc.*, **52**, 390.

[15] Johnston, R.M., 1973, *Color Theory*, in *Pigments Handbook, Vol. III*, ed. Patton, T.C., Wiley, New York.

[16] Barron, V. and Torrent, J., 1971, *J. Soil Sci.*, **37**, 499.

3. COLOUR AND TRICHROMATIC COLORIMETRY OF TRANSITION METAL COMPOUNDS

The most numerous group of coloured inorganic chemical compounds is that formed by compounds of the transition, i.e. d-electronic, elements. The valence electronic configuration of these elements is $(n+1)s^2nd^q$, where n $= 3$, 4, or 5 and q is the number of d-electrons $(0 \leq q \leq 10)$. The excitation energy of these electrons ranges from approximately 167 to 300 kJ, depending on the positions of excited states above the ground state. This range of energies corresponds to wavelengths from 714 to 400 nm, or wavenumbers from 14 000 to 25 000 cm^{-1}.

This narrow range of the electromagnetic spectrum is associated with a very large number of coloured compounds. Many thousands of these are transition metal compounds, the colours of a selection of which can be seen in Figures C4 to C8. The number and nature of such coloured species may be attributed to:

(a) the existence of numerous thermodynamically stable oxidation states, which correspond to different electronic configurations and states;

(b) various symmetries of coordination compounds of transition metals, which result in a variety of types of electronic spectra and hence various colours (hues);

(c) variation of strength of metal–ligand interaction (Dq).

Factor (b) is closely linked to the possibility of different coordination numbers, isomers, and generally different geometrical structures.

Colours of chemical substances have long been considered of prime importance. In days gone by colours were often used to identify compounds or isomers – thus $FeSO_4 \cdot 7H_2O$ was known as green vitriol, and $SnCl_4 \cdot 2NH_4Cl$, used in dyeing, as pink salt. The iron-cyanide complexes $K_3[Fe(CN)_6]$, $K_4[Fe(CN)_6]$, and $K_2[Fe(CN)_5(NO)]$ were known respectively as red, yellow, and black prussiates of potash, and iron-nitrosyl-sulphides as Roussin's red and black salts (now known to be alkali metal salts of $[Fe(NO)_2S]^-$ and of $[Fe_4(NO)_7S_3]^-$ respectively. The custom of naming by colour is not entirely dead today. Thus, for instance, the allotropes of phosphorus are often specified by their colour, though it should be added that the existence of

Table 3.1 Colours and names of coordination compounds

(a) **COBALT(III)**[a]

Colour	Historic name[b]	Systematic name	Formula
Yellow	luteocobalt chloride	hexaamminecobalt(III) chloride	$[Co(NH_3)_6]Cl_3$
Yellow brown	xanthocobalt chloride	nitropentaamminecobalt(III) chloride	$[Co(NH_3)_5(NO_2)]Cl_2$
Yellow	croceocobalt chloride	trans-dinitrotetraamminecobalt(III) chloride	trans-$[Co(NH_3)_4(NO_2)_2]Cl$
Brown	flavocobalt chloride	cis-dinitrotetraamminecobalt(III) chloride	cis-$[Co(NH_3)_4(NO_2)_2]Cl$
Purple red	chloropurpureocobalt chloride[c]	chloropentaamminecobalt(III) chloride	$[Co(NH_3)_5Cl]Cl_2$
Green	chloropraseocobalt chloride[c]	trans-dichlorotetraamminecobalt(III) chloride	trans-$[Co(NH_3)_4Cl_2]Cl$
Rose	roseocobalt chloride	aquapentaamminecobalt(III) chloride	$[Co(NH_3)_5(H_2O)]Cl_3$
Violet	chlorovioleocobalt chloride[c]	cis-dichlorotetraamminecobalt(III) chloride	cis-$[Co(NH_3)_4Cl_2]Cl$
Black	melano chloride	(mixture of three binuclear complexes)	$[(H_3N)_3Cl_2CoNH_2Co(NH_3)_3Cl_2]Cl$ etc.

(b) **PLATINUM**

Colour	Historic name	Formula
Green	Magnus's green salt	$[Pt(NH_3)_4][PtCl_4]$
Pink	Magnus's pink salt	$[Pt(NH_3)_3Cl]_2[PtCl_4]$
Red	Wolffram's red salt	$[Pt^{II}(EtNH_2)_4][Pt^{II}(EtNH_2)_4Cl_2]Cl_4$

[a] Analogous names were sometimes used for complexes of other metals, e.g. luteochromium(III) chloride for $[Cr(NH_3)_6]Cl_3$

[b] The colour prefixes are related to the following Latin words:
croceus (adj) – yellow, golden, saffron-coloured
flavus (adj) – yellow or gold-coloured; flavere (v) – to be yellow
luteus (adj) – yellow, orange, gold, saffron
melania (n) – blackness
prasius (n) – leek-coloured precious stone; prasinius (adj) – leek-green, greenish
purpureus (adj) – purple, violet, reddish, brownish
roseus (adj) – rosy, rose-coloured
viola (n) – violet; violaceus (adj) – violet-coloured
xanthos (n) – golden-coloured precious stone
The multiple meanings of several of these words add a further element of imprecision!

[c] The chloro prefix was often omitted; chloro could be replaced by bromo for the respective bromide-containing complexes.

two forms of white phosphorus and several forms that are more or less red in colour detracts from the usefulness of colour classification here. Colour nomenclature also surfaces in inorganic biochemistry, witness for instance purple acid phosphatase and cytochrome c_{550}, christened from its λ_{max} value.

The colours of coordination compounds containing transition elements have long been the object of chemists' interest. Indeed in the days before structures were readily established a number of complexes were identified by their colours. This approach was for many years, from the time of Frémy [1][1], reflected in the use of colour prefixes, particularly in naming cobalt(III) complexes (several examples are given in Table 3.1) [2]. Some platinum complexes were identified both by colour and discoverer (Table 3.1, bottom half). However, it is clear that qualitative characterisation by such names cannot be sufficiently informative and precise for the identifying these compounds. Thus colours of members of, e.g., the purpureo, $[Co(NH_3)_5X]X_2$, or croceo, trans-$[Co(diamine)_2X_2]X$, series vary significantly with the nature of X. Also whereas cis and trans isomers often differ in colour,[2] strikingly in the case of $[Co(en)_2Cl_2]^+$ (purple cis vs. green trans), much less dramatically for $[Cu(glycinate)_2]$ (blue cis vs. violet trans), both cis and trans isomers of $[Co(en)_2(NH_3)(NCS)](NCS)_2$ are essentially identical in colour. As has been said above, only quantitative measurement, under strictly defined experimental conditions, can be the basis for the use of colour in characterising compounds more precisely. This issue will be discussed in the remainder of this book.

Apart from the factors named above, one should mention the role of unstable oxidation states of transition elements, which may occur in the course of various reactions, especially in oxidation and reduction reactions. If it is assumed that the most likely redox reactions are single-electron reactions, it can be expected that in reactions of compounds involving high oxidation states of the metal, the occurrence of compounds involving lower states (thermodynamically unstable) may lead to faster or slower changes in the colour of the system, especially if the reaction takes place in solution, or even in the solid phase.

As was said in Chapter 2, the principal types of electronic transitions that may be the source of colour in transition metal compounds are ligand field and charge-transfer transitions.

3.1 COLOUR OF IONS AND COMPOUNDS OF TRANSITION METALS CAUSED BY d-d LIGAND FIELD TRANSITIONS

Ligand field transitions are associated with the presence of d electrons in the valence shell of the transition element in its compounds. We shall now briefly review the main issues related to this.

---

[1] It should be noted that the Polish chemist Józef Rogójski, about whom Zawidzki wrote in *Chemik Polski*, Vol. 11, 1911, dealt with 'luteocobaltiacks', i.e. complexes of cobalt(III) and ammonia (Prof. R. Sołoniewicz, personal communication).
[2] Thirteen examples are tabulated on page 1414 of N.V. Sidgwick, *The Chemical Elements and Their Compounds*, Vol. 2, Clarendon Press, Oxford, 1950.

3.1.1   Electronic states and configurations of atoms and ions of d-electronic elements

Table 3.2 shows the electronic configurations of the atoms of the elements under consideration in their ground states. The electronic configurations of ions in a given oxidation state are generally assumed to reveal only the presence of d-electrons. This is understandable for M^{2+} ions, as the electrons $(n + 1)\, s$ (for $n = 3, 4, 5$) are the first to undergo ionisation. But transition elements can occur in various oxidation states, including ions in formally negative states. Examples of transition metal oxidation states are given in Table 3.3.

The ligand field transitions that we discuss here are d-d type transitions, and the excitation of the d electron occurs within the d subshell. The most common geometric structure in complex transition metal compounds is the octahedral structure with the coordination number 6 and the tetrahedral or square planar structure with the coordination number 4. Examples of these and other structures, with corresponding coordination numbers (CN), are given in Table 3.4.

According to crystal field theory (CFT), in the ligand field surrounding the central ion of a transition metal d^1 with octahedral symmetry O_h, a splitting of the energy level D takes place into two sublevels denoted by the group theoretical symbols T_{2g} and E_g. The splitting enables the excitation of an electron (as the effect of electromagnetic radiation) from the ground state to the excited state. If the excitation

Table 3.2 Electronic configurations of transition elements atoms (in the ground state)

d	3	4	5	6	7	8	9	10	11	12
P										
Z 4	^{21}Sc $3d^14s^2$	^{22}Ti $3d^24s^2$	^{23}V $3d^34s^2$	^{24}Cr $3d^54s^1$	^{25}Mn $3d^54s^2$	^{26}Fe $3d^64s^2$	^{27}Co $3d^74s^2$	^{28}Ni $3d^84s^2$	^{29}Cu $3d^{10}4s^1$	^{30}Zn $3d^{10}4s^2$
Z 5	^{39}Y $4d^15s^2$	^{40}Zr $4d^25s^2$	^{41}Nb $4d^45s^1$	^{42}Mo $4d^55s^1$	^{43}Tc $4d^65s^1$	^{44}Ru $4d^75s^1$	^{45}Rh $4d^85s^1$	^{46}Pd $4d^{10}$	^{47}Ag $4d^{10}5s^1$	^{48}Cd $4d^{10}5s^2$
Z 6	^{57}La $5d^16s^2$	^{72}Hf $4d^26s^2$	^{73}Ta $5d^36s^2$	^{74}W $5d^46s^2$	^{75}Re $5d^56s^2$	^{76}Os $5d^66s^2$	^{77}Ir $5d^76s^2$	^{78}Pt $5d^96s^1$	^{79}Au $5d^{10}6s^1$	^{80}Hg $5d^{10}6s^2$

Table 3.3 Oxidation states and coordination numbers CN of transition elements[a]

Oxidation state	CN
−3	4, 5
−2	4, 5
−1	6,
0	6,
+1	6, (2), (3), (4)
+2	4, 6, 5, 7, 8, (12), (10)
+3	3, 4, 5, 6, (7), 8
+4	4, 5, 6, (7), 8, (11), (12)
+5	4, 5, 6, (7), 8
+6	4, 5, 6
+7	4, 6, (7), 5, (9)
+8	4, 6

[a] CN given in brackets are less common.

Table 3.4 Some examples of coordination geometry of d-block transition metal compounds and complexes

Coordination number	Geometry	Examples
4	Tetrahedral	$Ni(CO)_4$; $TiCl_4$; $[VO_4]^{3-}$; $[FeCl_4]^-$
	Square planar	$[Ni(CN)_4]^{2-}$; $[Pt(NH_3)_4]^{2+}$
5	Trigonal bipyramidal	$MoCl_5$; $[CuCl_5]^{3-}$
	Square pyramidal	$[VO(acac)_2]$; $[InCl_5]^{2-}$
6	Octahedral	$[Cr(H_2O)_6]^{3+}$; $[Co(NH_3)_6]^{3+}$; $[PtF_6]^{2-}$
8[a]	Square antiprismatic	$[[TaF_8]^{3-}$; $[Zr(acac)_4]$
	Dodecahedral	$[Mo(CN)_8]^{3-}$

[a] Cube geometry is known for a few eight-coordinate f-block complexes, for example $[U(NCS)_8]^{4-}$ (in R_4N^+ salts) and $[PaF_8]^{3-}$.

corresponds to wavelengths in the visible range, the absorption spectrum contains a band (a maximum), and the part of the radiation which is transmitted through the compound results in an appropriate colour – as has already been described in detail.

To get a fuller picture of this problem, it should be remembered that in the example above only the d^1 electronic configuration was discussed, such as occurs in the compounds of Ti(III) and Mn(VI). Of course, a multielectron configuration causes the occurrence of a large number of terms. Let us remember that for the configuration d^q, that is a configuration of equivalent electrons ($n = $ const, $l = $ const $ = 2$), the following terms occur (the ground term is underlined):

$$d^1, d^9 \quad {}^2\underline{D}$$
$$d^2, d^8 \quad {}^1S\ {}^1D\ {}^1G\ {}^3P\ {}^3\underline{F}$$
$$d^3, d^7 \quad {}^2P\ {}^2D(2)\ {}^2F\ {}^2G\ {}^2H\ {}^4P\ {}^4\underline{F}$$
$$d^4, d^6 \quad {}^1S(2)\ {}^1D(2)\ {}^1F\ {}^1G(2)\ {}^1H\ {}^3P(2)\ {}^3D\ {}^3F(2)\ {}^3G\ {}^3H\ {}^5\underline{D}$$
$$d^5 \quad\quad {}^2S\ {}^2P\ {}^2D(3)\ {}^2F(2)\ {}^2G(2)\ {}^2H\ {}^2I\ {}^4P\ {}^4D\ {}^4F\ {}^4G\ {}^6\underline{S}$$

These terms are the so-called atomic terms. Generally the term for a given electronic configuration is represented by the notation ${}^{2S+1}L_{(J)}$, where S denotes the total spin quantum number, $2S + 1$ is the so-called multiplicity, L stands for the total orbital quantum number, and J is the total internal quantum number.

The description of the energetic state of d^q electronic ions must be extended in the case of actual, simple or complex, transition metal compounds by recognising the effect of the environment of the ion. If we represent a transition compound in the most general way as M_yL_x, where M denotes the metal ion and L the ligand, then in the simplest case $y = 1$ defines a mononuclear complex with the coordination number equal to x (on the assumption that L is a monodentate ligand, i.e. with a single-atom bond between the ligand and metal ion). As has been concluded above for the d^1 case, the electrostatic field created by the ligand causes a change in the energy levels, namely either their splitting or a change in the energy value of the levels which have not been split. This also applies to multielectron configurations. What this means is that depending on the symmetry of the environment of the ion, there will be a certain number of energy levels. New terms are denoted by different group theoretical symbols.

In the previously described situation of the d^1 system, the five-fold degenerate term ^2D ($2L + 1 = 5$) is split into two terms, $^2T_{2g}$ and ^2E_g. Generally in O_h symmetry there may be the following terms: A_1, A_2, E, T_1, T_2, the first two of which are not degenerate, the third is two-fold degenerate, and the rest are threefold degenerate.

3.1.2 Structure, symmetry, and colour

Considering the question of the nature of colour and its relationship to the electronic spectrum, we note that the simplest case involves only the electron configurations d^1 and d^9, which have just one absorption band. For other electron configurations (so-called high-spin configurations), the number of (spin-allowed) absorption bands for octahedral symmetry is 3. In such cases it is in principle not possible to predict precisely the colour of a compound on the basis of the position of those bands. Besides, depending on oxidation state and ligand type, the splitting of energy levels (expressed by the parameter Dq, to which we shall return shortly) is liable to change and a substantial increase or decrease of the splitting may cause the electronic spectrum to move towards the ultraviolet (hypsochromic shift) or towards the infrared (bathochromic shift)[3] As a result, only two bands, or even one, may be observable in the visible range. The position of the absorption minimum or minima, i.e. the regions of light transmission, provides important qualitative information. Thus, for instance, the absorption spectrum of $[Ni(H_2O)_6]^{2+}$ reveals two absorption bands and a deep minimum in the visible range (at approximately 500 nm), which clearly suggests that the compound should show green colouration.

The magnitude of splitting for one d electron in a field with O_h symmetry, between the terms $^2T_{2g}$ and ^2E_g, is defined as $10Dq$. It is a semiempirical parameter (which in principle, under certain assumptions, can be calculated theoretically) which is the fundamental quantity in crystal field theory. This subject has been extensively described in literature on electronic spectroscopy (see, e.g. [3–8]) and will not be considered here in any detail.

Table 3.5 presents the energy values expressed as Dq values for terms in O_h symmetry (for terms with maximum multiplicity).

Thus, if the Dq value is known, it is possible to calculate the excitation energy of the electrons and to predict the approximate position of absorption bands. Consequently it is also possible to predict partially the transmittance range and hence the likely colour (Figure 2.1).

Such reasoning is to a large extent an approximation. First, the energy of energy levels in multielectron configurations also depends on interelectronic interaction, which is expressed in the form of the so-called Condon-Shortley interaction integrals or Racah parameters B and C. As a result, the energy is a function of Dq, B, and C (for octahedral complexes). Second, colour depends on the envelope of a concrete absorption spectrum, as has been mentioned above.

---

[3] Moreover, intense charge-transfer absorption may extend into the visible region, masking one or more of the weak d-d bands and exerting a dominant effect on the colour.

Table 3.5 Energy of terms with highest multiplicity for d^q electronic configuration in weak crystal fields of O_h symmetry

Electronic configuration	Atomic term	Energy, Dq			
d^1	^2D	$^2T_{2g}(-4)$	$^2E_g(+6)$		
d^2	^3F	$^3T_{1g}(-6)$	$^3T_{2g}(+2)$	$^3A_{2g}(+12)$	
d^3	^4F	$^4A_{2g}(-12)$	$^4T_{2g}(-2)$	$^4T_{1g}(+6)$	
d^4	^5D	$^5E_g(-6)$	$^5T_{2g}(+4)$		
d^5	^6S	$^6A_{1g}(0)$			
d^6	^5D	$^5T_{2g}(-4)$	$^5E_g(+6)$		
d^7	^4F	$^4T_{1g}(-6)$	$^4T_{2g}(+2)$	$^4A_{2g}(+12)$	
d^8	^3F	$^3A_{2g}(-12)$	$^3T_{2g}(-2)$	$^3T_{1g}(+6)$	
d^9	^2D	$^2E_g(-6)$	$^2T_{2g}(+4)$		

It turns out that the Dq value can be estimated on the basis of the approximate parametrisation $Dq = f \cdot g$, where g is the characteristic value in cm^{-1} for the central ion (in a given oxidation state) and f is a dimensionless quantity for the given ligand. These data for many transition metals and ligands are to be found in the literature we have already cited above (e.g. [4, 5, 7]). Such parametrisation has been used in simulations of absorption spectra and calculations of chromaticity coordinates for certain coordination compounds (as will be discussed below). Transition metal compounds often manifest a deformation of octahedral structures, for instance as a result of a change of the distance of some ligands from the central ion. Such deformations are mostly found in mixed complexes whose composition can be formally expressed as $ML_x^1 L_z^2$, where $x + z = $ coordination number of the complex (L^1 and L^2 are unidentate ligands). They are mostly complexes with D_{4h} or C_{4v} (tetragonal) symmetry. This lower symmetry causes a further splitting of energy levels, which in principle leads to the formation of a larger number of absorption bands in the spectrum of the compound. However, such bands are often not visible or show up as a shoulder or inflection and computer resolution methods are needed to isolate them. These issues are discussed in detail in [9].

Of course it is not necessary to know the exact values of transition energies in order to determine colour and chromaticity coordinates as quantitative calculations of colour are done on the basis of a concrete (optical) electronic spectrum envelope. Yet, in chemistry the observed colour is often ascribed to a specific form of, for instance, a complex compound. Such 'identification' may lead to false conclusions.

3.1.2.1 Spectrochemical series and colour

Before we proceed to discuss in more detail the problems which arise when dealing with the distortion of octahedral complexes frequently encountered in practice, or with mixed complexes, we will stress the role of the Spectrochemical Series. As emphasised previously, the treatment of the energetics of d^q ions must be extended in the case of actual transition metal compounds, simple or complex, by considering the overall effect of the environment of the ion. In most cases the nearest neighbours, in other words the (first) coordination sphere of ligands, are of paramount importance. These ligands cause a splitting of the energy levels of the metal (central) ion to an extent depending, among other factors, on the symmetry of the whole system. As a consequence the electronic spectrum of the complex is moved to shorter or longer wavelengths, depending on the ligand. We neglect, at this stage, other changes in the spectrum, such as band intensity and band half-width.

Comparison of $10Dq$ values for a set of complexes of a given metal ion with different ligands L leads to a so-called Spectrochemical Series. Historically, this idea was due to Fajans [10] and Tsuchida [11]. Subsequently this matter has been extensively studied, especially by Jørgensen [12]. The spectrochemical series is, to a first approximation, independent of the nature of the central metal ion, and shows the following ordering of increasing $10Dq$, for both octahedral or tetrahedral symmetries:

$$I^- < Br^- < CrO_4^{2-} < Cl^- \approx SCN^- < N_3^- \approx S_2O_3^{2-} < CO_3^{2-} < ONO^-$$
$$\approx OH^- < SO_4^{2-} < NO_3^- < O_2CCO_2^{2-} < H_2O < NCS^- < edta^{4-} < py$$
$$\approx NH_3 < en < SO_3^{2-} < bipy < phen < NO_2^- < CN^-$$

(SO_4^{2-}, NO_3^-, CO_3^{2-}, and $S_2O_3^{2-}$ here all acting as monodentate ligands).

Schematically we can illustrate this spectrochemical series of ligands as shown in Figure 3.1 and Figure 3.2. The former is simple to understand, as it shows the increase of the parameter $10Dq$ according to the increase in the values of the parameter f in Table 3.6 – for a constant value of g, i.e. the same metal (central) ion. Figure 3.2 is a little more complex. In contrast to Figure 3.1, which corresponds, for instance, to the d^1 electronic configuration (as well as to d^9), it presents the course of increasing values of wave numbers σ_i of the three spin-allowed bands which appear in the spectra of octahedral complexes of Ni^{2+}. The interesting point to be stressed is the non-linear trend for σ_2 and σ_3. As will be explained, this is caused by the interaction between two states having the same group theoretical symbols {here between $^3T_{1g}(F)$ and $^3T_{1g}(P)$}.

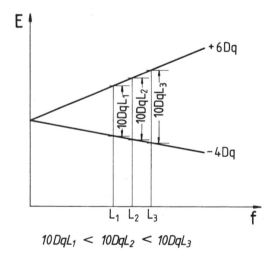

Figure 3.1 Schematic illustration of a spectrochemical series of ligands (L_1, L_2, L_3,...) as a function of f (at constant g).

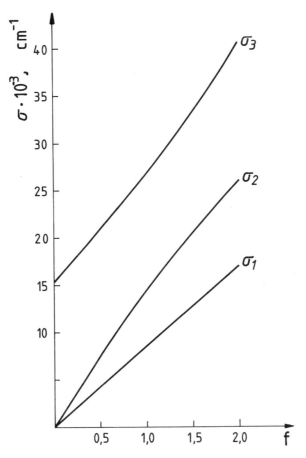

Figure 3.2 Dependence of wavenumbers σ_i for electronic transitions on f, using Ni^{2+} as example.

Table 3.6 Values of f and g for ligands and central metal ions in complexes

Ligands	f	Central ion	g (kK)
$6CN^-$	1.7	V(II)	12.3
3 en	1.28	Cr(III)	17.4
$6NH_3$	1.25	Mn(II)	8.0
$6H_2O$	1.00	Mn(IV)	23
$3C_2O_4^{2-}$	0.98	Fe(III)	14.0
$6CO(NH_2)_2$	0.91	Co(III)	19.0
$6F^-$	0.9	Ni(II)	8.9
$6Cl^-$	0.80	Mo(III)	24.0
$6Br^-$	0.76	Rh(III)	27.0
		Re(IV)	35
		Ir(III)	32
		Pt(IV)	36

The question now arises as to whether the spectrochemical series of ligands could be predictive for the colours of complexes, and how far such a prediction might be realistic. From the ordering of ligands as already given it may be concluded that the whole contour of the spectrum would be correspondingly shifted, hypsochromically or bathochromically. By the same token it follows that the absorption band positions are also shifted, and as a result the colour of the object must change. One main condition must be fulfilled, namely that the spectrum contour should not change significantly, since the colour characteristics are determined by this feature.

When considering the colour of complexes in relation to the spectrochemical series, one further point deserves attention. Several of the ligands generally included, such as oxalate, edta, bipy, and phen in the series reproduced above, are not monodentate. The geometry of a complex with, say, a bidentate ligand, is not strictly octahedral, and this may cause not only a different splitting of energy levels but also a change of colour of the compound under investigation. It is, however, generally assumed that all members of the spectrochemical series create octahedral, or at least pseudo-octahedral, symmetry and that complexes of different metal (central) ions can be interpreted in the same manner. It should be borne in mind that there are several exceptions, especially those complexes which are four- rather than six-coordinate. There is also the possibility, particularly for cyanide, bipy, and phen, that particularly strong ligands may cause a change in magnetic properties from high-spin to low-spin. The particular case of nickel(II) illustrates these points, for nickel(II) is found in square-planar and tetrahedral environments as well as in the more normal octahedral. Moreover the strong-field ligand cyanide causes both change in geometry, to square-planar, and change in magnetic properties, to low-spin. Hence here the colour change is jointly due to a high Dq value and different symmetry. We discuss some of these problems further on.

It is interesting to note that the O^{2-} and OH^- ions, and indeed water, commonly enough encountered ligands in transition metal chemistry, are generally not included in the spectrochemical series. However according to Jørgensen [13], Werner stated as early as 1912 that there was a regular hypsochromic shift for donor atoms: $I < Br < Cl < F < O < S \approx N < C$. As we shall see, the influence of the O^{2-} ligand on the colour of compounds is rather closely related to the metal-oxygen distance, as demonstrated mainly for oxides but also for oxoanions. Colour changes in the $Al_2O_3 + Cr_2O_3$ system, as dependent on the Dq parameter, are discussed in Chapter 7 (Figure 7.3). The role of metal ionic radii in colours of oxoanions and oxocations is mentioned again in Section 3.2.1 later in this Chapter.

Just as ligand protonation, especially the $O^{2-} \rightarrow OH^- \rightarrow H_2O$ sequence alluded to above, affects colour and Dq values, so also does metallation of a coordinated ligand such as cyanide [14] or thiocyanate [15]. The best known and studied case is addition of Hg^{2+} to thiocyanate coordinated to cobalt, as in $[Co(NH_3)_5(NCS)]^{2+}$ or $[Co(NCS)_4]^{2-}$. Addition of Hg^{2+} to a solution containing $[Co(NH_3)_5(NCS)]^{2+}$ causes a colour change from orange to yellow; addition of Hg^{2+} to a solution containing $[Co(NCS)_4]^{2-}$ causes a shift of λ_{max} from 624 to 604 nm (ν_{max} from 16 030 to 16 560 cm^{-1}). The metallated ligand $-NCSHg^+$ lies further towards the 'strong ligand' end of the spectrochemical series than $-NCS^-$ itself [16].

As previously stated, and depicted in Figure 3.3, in some d^q electronic configurations two or more energy levels do interact, leading to a deviation from the linear

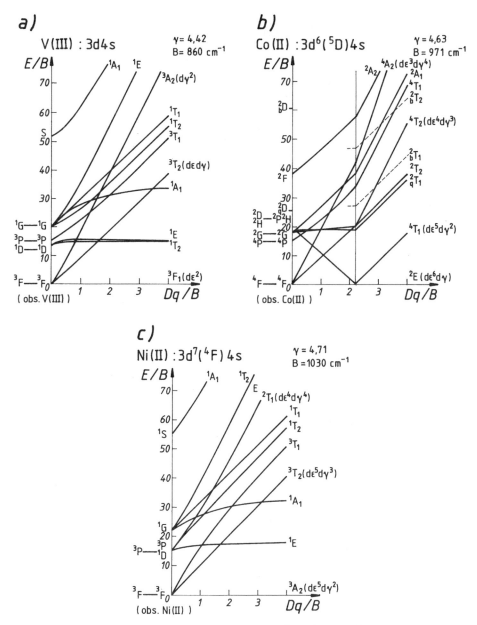

Figure 3.3 Tanabe-Sugano energy level diagrams for ions a) d^2, b) d^7, and c) d^8; $d\varepsilon$ denotes t_{2g} electrons, $d\gamma$ e_g electrons.

dependence given by the simple relation $E = f(Dq)$. As examples, the so-called Tanabe-Sugano diagrams, of E/B plotted against Dq/B, are shown for d^2, d^7, and d^8 configurations. Both axes incorporate Racah's B parameter, which has already been mentioned and will be further discussed in more detail in Section 3.1.2.2.

When looking at the Dq parameter from a chemical point of view, it was stated that $10Dq$ (or, as sometimes labelled, Δ) is approximately given by

$$10Dq \approx \text{electrostatic (first-order) perturbation} + \sigma(L \to M) + \pi(M \to L) - \pi(L \to M).$$

According to the angular overlap model (AOM),

$$10Dq = 3e_\sigma(L) - 4e_\pi(L).$$

As e_σ is always positive, when sigma-bonding is very small the position of a ligand in the spectrochemical series is determined by its π-bonding ability, as donor or as acceptor. The latter corresponds to a position to the right of the series; the former type of ligands, being π-donors, appear towards the left of the series set out above. These properties cause different splitting patterns of energy levels and thus also possibly a change in colour of complexes. As stated previously, the prediction of colour must be treated with care.

For a given ligand one should also expect to be able to construct a spectrochemical series for metal ions. This can be achieved in practice, with the following approximate ordering of metal ions:

$$\text{Mn}^{\text{II}} < \text{Co}^{\text{II}} \sim \text{Ni}^{\text{II}} < \text{V}^{\text{II}} < \text{Fe}^{\text{III}} < \text{Cr}^{\text{III}} \sim \text{V}^{\text{III}} < \text{Co}^{\text{III}} < \text{Mn}^{\text{IV}} < \text{Mo}^{\text{III}}$$
$$< \text{Rh}^{\text{III}} < \text{Ir}^{\text{III}} < \text{Pt}^{\text{IV}}.$$

There are two main empirical relations to be applied to this series. Values for $10Dq$ increase in the orders:

(1) $\text{M}^{\text{II}} < \text{M}^{\text{III}} < \text{M}^{\text{IV}}$

(2) $3d < 4d < 5d$

It seems that the spectrochemical series for metal ions (with the same ligand) would, like the ligand series, not provide firm information for colour prediction. However, if colour is due to charge-transfer transitions, as in the transition metal d^0 oxoanions, the second ordering can be successfully applied. This will be discussed in Section 3.2.

As has been stated, real complex compounds frequently deviate, for a number of reasons, from perfect octahedral, or perfect tetrahedral, symmetry. In lower symmetry fields, of which D_{4h}, C_{4v}, D_{2d}, and C_{2v} are commonly encountered, the number of energy levels increases and the number of electronic transitions and bands therefore is likely to increase. Figures 3.4 and 3.5 illustrate this.

For tetragonal (D_{4h}) complexes two parameters, Ds and Dt, are needed to supplement Dq in order to express the energies of the extra levels. In the case of a d^1 electronic configuration, these energies are as follows:

$$E(b_{1g}) = 2Ds - Dt \qquad E(b_{2g}) = 2Ds - Dt$$
$$E(a_{1g}) = -2Ds - 6Dt \qquad E(e_g) = -Ds + 4Dt.$$

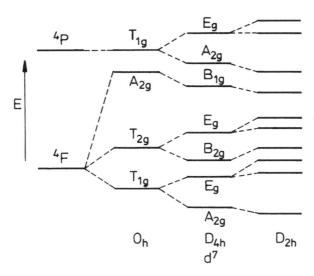

Figure 3.4 Energy level splitting in ligand fields of O_h and successively lower symmetries for the d^7 electronic configuration.

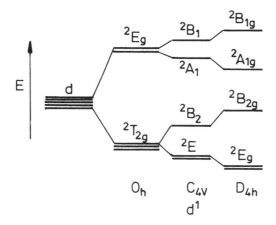

Figure 3.5 Energy level splitting in ligand fields of O_h and successively lower symmetries for the d^1 electronic configuration.

3.1.2.2 *The nephelauxetic series and colour*

As stated by Jørgensen [7] 'the nephelauxetic effect is that the phenomenological parameters of interelectronic repulsion are smaller in complexes than in the corresponding gaseous ions'. Interelectronic repulsion has been treated by Condon and Shortley [17] and by Racah [18]. According to the latter, two such quantities should be formulated for d electrons, viz. B and C. It is frequently assumed that $C \sim 4B$, although experimental data obtained from the analysis of absorption spectra of complexes depart considerably from this assumption in many cases. According to

Jørgensen's dictum above, B and C are reduced in relation to their values in the gas phase, and a parameter is defined as the ratio of B_k in the complex to B_g for the free ion. Collecting such ratios for a wide range of complexes we obtain the following Nephelauxetic Series based on B values for ligands and for metal ions:

$$F^- > H_2O > (CH_3)_2SO \approx CH_3CONH_2 > (NH_2)_2CO > NH_3 > C_2O_4^{2-} \approx$$

$$H_2NCH_2CH_2NH_2 > Cl^- > CN^- > Br^- > (C_2H_5O)PS_2^- \sim S^{2-} \sim I^- >$$

$$(C_2H_5)_2NCS_2^- > (C_2H_5O)_2PSe_2^-$$

$$Mn^{II} \sim V^{II} > Ni^{II} > Co^{II} > Mo^{II} > Fe^{IV} \sim Cr^{III} > Fe^{III} > Os^{IV} > Ir^{III} >$$

$$Rh^{III} > Co^{III} > Pt^{IV} \sim Mn^{IV} > Ir^{VI} > Pt^{VI}$$

It is interesting to examine the spectroscopic results of the nephelauxetic effect. Relative to the gas phase there is a bathochromic shift of corresponding bands. In lanthanide complexes this shift is rather, or even very, small (around 1%), but in d-block compounds the effect may be considerable. As shown by Jørgensen [19], three nephelauxetic parameters, denoted β_{55}, β_{35}, and β_{33} should be formulated, labelled according to which orbital (electronic cloud) is expanded. The subscripts to the β parameters relate to his use of the labels γ_5 and γ_3 for t_2 and e orbitals. These nephelauxetic parameters are defined as:

$$\beta_{55} = B_k/B_g \qquad \beta_{35} = B_k/B_g(1-\varepsilon) \qquad \beta_{33} = B_k/B_g(1-\varepsilon)^2,$$

where ε is a covalency parameter. These formulae may be explained by molecular orbital theory, as described by Schäffer and Jørgensen [20], and will not be treated in detail in this book.

There are some trends which may be formulated in the nephelauxetic series. Firstly, if the series is formulated in terms of $(1 - \beta)$ instead of β, then it can be factorized into functions of ligand only and of metal ion only:

$$(1 - \beta) = h \text{ (ligands)} \times k \text{ (metal ions)}.$$

Values for h and k are listed in Table 3.7. It is evident that k is strongly dependent on the nature of the central metal ion. For instance, k for V^{II} is 0.08, but for Pt^{IV} k is six times larger, at 0.5. From the fluoride octahedral values for these two ions given in Table 3.7 it is clear that they differ markedly in their B_k values, which are $380\,cm^{-1}$ for Pt^{IV}, $690\,cm^{-1}$ for V^{II}. This corresponds to a considerable increase in covalent bonding in Pt^{IV} complexes compared with V^{II}. Generally speaking, the tendency to increasing covalency is more particularly evident in the nephelauxetic series of ligands as compared with the spectrochemical series of ligands.

When ordering ligands according to ligating atoms:

$$F^- < O^{2-} \text{ (from } H_2O) < Cl^- < Br^- < I^- < O^{2-} \text{ (in oxides)} < S^{2-},$$

it becomes clear that there is a correlation with ligand reducing power. As in the case of the spectrochemical series, the position of O^{2-} (oxide; water) is uncertain. The role

Table 3.7 Values of h and k for ligands and central metal ions in complexes

Ligands	h	Central ion	k
6F$^-$	0.8	V(II)	0.08
6H$_2$O	1.0	Cr(III)	0.21
6CO(NH$_2$)$_2$	1.2	Mn(II)	0.07
6NH$_3$	1.4	Mn(IV)	0.5
3en	1.5	Fe(III)	0.24
3C$_2$O$_4^{2-}$	1.5	Co(III)	0.35
6Cl$^-$	2.0	Ni(II)	0.12
6CN$^-$	2.0	Mo(III)	0.15
6Br$^-$	2.3	Rh(III)	0.30
		Re(IV)	0.2
		Ir(III)	0.3
		Pt(IV)	0.5

of bond length is relevant to this problem with the nephelauxetic series – the metal-oxygen bond length is much longer for aqua-complexes than for oxide complexes.

It is difficult to draw any firm conclusions about the role of B values in the colours of complexes. Generally speaking, an increase in covalent bonding in the complex, and hence a decrease in B, should cause a bathochromic shift and consequent change in colour. Pauling [21] showed how the colours of sulphides, iodides, bromides, and chlorides of different metals depended on the extent of covalent bonding, expressed in terms of formation enthalpies (Table 3.8). The smaller the enthalpy, and the greater the electronegativity of the element, the greater the red shift and the greater the intensity of the colour – in the extreme to the compound appearing black. These observations on colour changes are, however, not simple to interpret, since the changes are caused simultaneously by a shift in the spectrum and an increase in intensity. It is difficult to decide which factor is the more important in determining the colour. Some data on these factors have been obtained as a result of our studies on simulation spectra and colour chromaticity coordinates (see Table 3.28 in Section 3.3).

Table 3.8 Colours of some halides and sulphides as dependent on bond covalency, expressed in terms of formation enthalpies in kcal mol^{-1}. The numbers in brackets are electronegativities

	S (2.5)	I (2.5)	Br (2.8)	Cl (3.0)
Na^I(0.9)	69	86	98	136
Mg^{II}(1.2)	42	43	62	77
Al^{III}(1.5)	20	25	42	55
Cd^{II}(1.7)	17 yellow	24 yellow	38	47
Pb^{II}(1.8)	11 black	20 yellow	33	43
Sn^{II}(1.8)	6 brown	19 yellow	31	41
Ag^I(1.9)	4 black	15 yellow	24 yellow	30
Sb^{III}(1.9)	7 orange	8 red	21 yellow	30
As^{III}(2.0)	6 red yellow	5 red	16 yellow	27
Pt^{II}(2.2)	7 black	5 brown	10 brown	16 red

3.1.2.3 Mixed complexes and isomers

An approximate method of evaluating the effective Dq value has been proposed for mixed complexes (octahedral and tetrahedral). It is the so-called mean environment rule given by the following equation (for octahedral complexes):

$$Dq\left(ML_x^1 L_{6-x}^2\right) = \frac{x}{6} Dq\left(ML_6^1\right) + \frac{6-x}{6} Dq\left(ML_6^2\right),$$

where $Dq(ML_6^1)$ is the Dq value for an octahedral complex with ML_6^1 composition, and $Dq(ML_6^2)$ is the Dq value for a complex with the second ligand. This method has been used to simulate the optical spectra and calculate the chromaticity coordinates of, among others, the Cr^{3+}-NH_3-H_2O system and the analogous system for Ni^{2+}.

For the equation $10Dq = f \cdot g(cm^{-1})$ the spectrochemical series of ligands is obtained (for a specific symmetry, normally octahedral) for variable f at constant g. Complementarily, varying g at constant f gives the spectrochemical series of metal (central) ions. Values of f and g for a range of metal ions and ligands have been given in Table 3.6, near the start of Section 3.1.2.1 above.

Generally the prediction of the colour of a compound for a given metal ion with changing ligands and unchanging symmetry on the basis of the spectrochemical series is straightforward. Increasing f values for the ligand suggest that generally the absorption bands will undergo a hypsochromic shift, while decreasing values indicate a bathochromic shift. The colour of a complex compound may change accordingly. If however there is a simultaneous significant change of the spectral envelope, for instance due to the increased intensity of an electronic transition or because of a drop in the number of bands in the visible spectrum (as has already been mentioned above), such prediction is impossible.[4]

The problem is more complicated for mixed complexes. If tetragonal symmetry is assumed, a larger number of absorption bands should be expected, which is not always to be seen in the experimental spectrum. Moreover, the parameter Dq is not the only parameter (in addition to B and C) for these systems which determine the energy of the split levels (as compared with the octahedral symmetry of a complex with a uniform coordination sphere). Two additional parameters must be introduced, Ds and Dt, whose values determine the strength of the crystal field as well as the deformation in the equatorial and axial planes of the complex.

The molecular formula of a mixed complex does not give an indication of the particular geometry of the compound. Yet, depending on the axial and planar strength of the ligand field, the absorption spectrum may be completely different, and consequently the colour of different isomers may be completely different.

Some effect on the colour of the given isomer may be expected for all types of isomerism. Let us recall that the following types are distinguished in coordination chemistry: geometric, ionization, structural, coordination, and polymerization

---

[4] These simple considerations do not take into account shifts from weak to strong crystal fields which can be caused by changing the ligand.

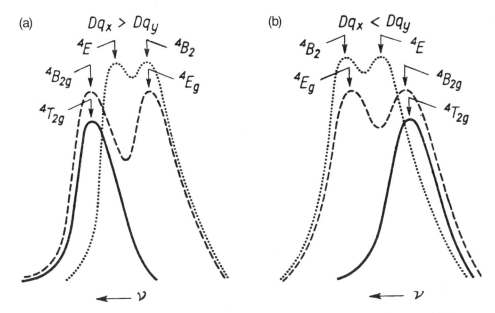

Figure 3.6 The dependence of quartet terms in *cis* and *trans* complexes on Dq_x and Dq_y; —— MX_6, - - - - *trans*-MX_4Y_2, ... *cis*-MX_4Y_2 (from ref. [111]).

isomerism. We shall try to see whether the direction of changes affecting the colour of different isomers can be predicted.

In practice, geometric isomerism occurs as *cis-trans* isomerism. For the MX_4Y_2 complex, bathochromic or hypsochromic shift was observed depending on the value of Dq_x in relation to Dq_y. Thus, if $Dq_x > Dq_y$, case a obtains; otherwise, case b (in Figure 3.6). This is accompanied by a change of colour, which can however be evaluated only in approximate terms.

For tetragonal complexes, the angular overlap model is particularly valuable and useful. In that model, it is possible to determine precisely the role of individual ligands by calculating the parameters e_σ and e_π, the latter corresponding to the contribution of the π bond, which crystal field theory neglects. These values have been calculated [22] for *trans*-[Ni(py)$_4$X$_2$], where py = pyridine and X = NCS$^-$, Cl$^-$, Br$^-$, and I$^-$. Figure 3.7 presents the spectra of these complexes.

It can be seen in Figure 3.7 that only the iodide complex stands out from the others, chiefly in the third absorption band, whose wavenumber is 28 609 cm^{-1} (calculated from crystal field theory), while for the chloride, bromide, and thiocyanate, the corresponding values are, respectively, 26 748, 26 114, and 27 284 cm^{-1}. What is of particular significance is the clearly marked change of strong light transmission in the green range, which is characteristic of many Ni(II) compounds. In the iodide complex, at the wavenumber of approximately 19 000 cm^{-1}, there is a substantial increase of absorption, in contrast to the usual minimum in this range.

In this complex there is a sharp decrease of the M–I σ bond strength, and a particularly sharp decrease of the π bond strength, with a simultaneous sharp change of the tetragonal deformation parameter Ds and a somewhat less sharp change of Dt.

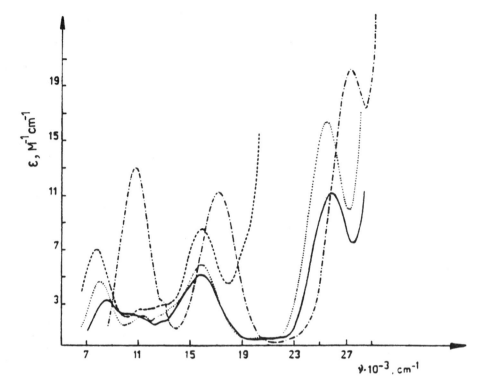

Figure 3.7 Absorption spectra of *trans*-[Ni(py)$_4$X$_2$] complexes in chloroform (at room temperature) [22]; —— X = Cl, – – – X = Br, ······ X = I, – · – · – X = NCS.

The data obtained from computer analysis which characterise tetragonal deformations in *trans*- and *cis*-[CrCl$_2$(H$_2$O)$_4$]Cl compounds are presented in Tables 3.9 and 3.10.

For practical purposes, we can apply the same line of reasoning to other types of isomerism. Thus, for instance, for ionization isomers of the type $M^{2+}L_4A_2$ and $M^{2+}L_5A$, the *change of coordination sphere* will have a significant effect on the colour if A and L differ substantially in field strength (i.e. if they are relatively far apart in the spectrochemical series.

3.1.2.4   Coordination number and stereochemistry

The colour changes of coordination compounds (for the same transition metal ion) are very characteristic and important in so-called coordination and configuration equilibria. The former comprise changes of the coordination number, for instance in reactions of the octahedral → tetrahedral type, which may be represented in the simplest form as $ML_6 \rightarrow ML_4 + 2L$, or octahedral → square planar (which corresponds to the same equation). Configuration equilibria occur if the coordination number is constant and the geometry varies, for instance trigonal bipyramidal – square pyramidal or tetrahedral – square planar.

Table 3.9 Band position, assignment, calculated parameters and transition energies for the complex trans-[CrCl$_2$(H$_2$O)$_4$]Cl, cm^{-1} [32]

Exp.	Resolution	Assignment	Calculated, CFM	Calculated, AOM
	14127	$^2A_{1g}(G, E_g)$	14329	14331
	14155	$^2B_{1g}(G)$	14350	14353
14600sh	14443	$^2E_g(G, T_{1g})$	14935	14938
	15080	$^2A_2(G)$	14942	14944
15500	15532	$^4E_g(F, T_{2g})$	15609	15611
15700	16206	$^4B_{2g}$	16509	16501
	21042	$^2B_{2g}(G)$	21404	21407
	21074	$^2E_g(G, T_{2g})$	21440	21443
22350	22284	$^4A_{2g}(F)$	22109	22102
	23134	$^4E_g(F, T_{1g})$	22836	22838
	28122	$^2A_{1g}(G, A_g)$	27938	27941
	29520	$^2E_g(H, T_{2g})$	29728	29732
	30520	$^2E_g(H, T_{1g}(2))$	30471	30476
	30615	$^2B_{2g}(H)$	30600	30595
	30951	$^2A_{2g}(H, T_{1g}(2))$	30920	30914
	33003	$^2A_{1g}(D(1))$	32212	32217
	32682	$^2B_{1g}(D(1))$	32304	32300
	35099	$^2A_{2g}(H, T_{1g}(1))$	34919	34914
	34870	$^4A_{2g}(P)$	35247	35251
	34901	$^4E_g(P)$	35376	35371
	35572	$^2E_g(H, T_{1g}(1))$	35934	35398
		r.m.s.	317	317
		Dq_{xy}	1651(10)	
		Dq_z	1477	
		Ds	−151(23)	
		Dt	−100(11)	
		B	675(16)	675(18)
		C	3128(33)	3128(31)
		C/B	4.6	4.6
		$e_\sigma(H_2O)$		6310(33)
		$e_\pi(H_2O)$		607(26)
		$e_\sigma(Cl)$		5759(17)
		$e_\pi(Cl)$		624(18)

Let us consider the problem of octahedral-tetrahedral equilibrium in solutions. The expectation of a change of colour is justified by the fact that formally the splitting of energy levels in tetrahedral symmetry is much smaller than in octahedral symmetry (for the same ligand and metal ion and the same M-L distance). Roughly, Dq (tet) = −4/9 Dq (oct) (the minus sign means that there is a reversal of energy levels). This means that the absorption spectrum of a tetrahedral complex undergoes a substantial bathochromic shift. A classic example, which will be discussed in detail in connection with the question of solvatochromism, is CoCl$_2$, which is pink in aqueous solutions. An increase of chloride ion concentration, introduced as HCl or a metal chloride, results in a change of the colour to deep blue, which can be explained by the following reaction:

$$[Co(H_2O)_6]^{2+} + 4Cl^- \rightarrow [CoCl_4]^{2-} + 6H_2O$$
pink   blue

Table 3.10 Band positions, assignment, calculated parameters and transition energies for the complex cis-$[CrCl_2(H_2O)_4]Cl$, cm^{-1} [32]

Exp.	Resolution	Assignment	Calculated, AOM
14500	14806	$^4A_1$	13559
	14149	$^2A_2(1)$	14666
	14299	$^2A_1(1)$	14701
	15244	$^2B_1(1) \cdot {}^2B_2(1)$	15006
	15253	$^2A_2(2)$	15075
15900	16183	$^4B_1(1) \cdot {}^4B_2(1)$	16687
22300	22201	$^4B_1(2) \cdot {}^4B_2(2)$	21638
	20688	$^2A_2(2)$	21951
	22118	$^2B_1(2) \cdot {}^2B_2(2)$	21953
	22826	$^4A_2(1)$	23160
	27711	$^2A_1(3)$	27631
	28064	$^2A_1(4)$	28362
	28693	$^2A_2(3)$	28683
30000	29899	$^2B_1(3) \cdot {}^2B_2(3)$	30025
	32062	$^2A_2(4)$	31666
	33288	$^2B_1(4) \cdot {}^2B_2(4)$	32691
	34633	$^2A_1(5)$	34277
	–	$^2B_1(5) \cdot {}^2B_2(5)$	34407
	34019	$^4B_1(3) \cdot {}^4B_2(3)$	34410
	34818	$^4A_2(3)$	35261

		r.m.s.	506
		$e_\sigma(H_2O)$	8010(81)
		$e_\pi(H_2O)$	772(55)
		$e_\sigma(Cl)$	3996(88)
		$e_\pi(Cl)$	1528(80)
		B	534(32)
		C	3542(65)
		C/B	6.6

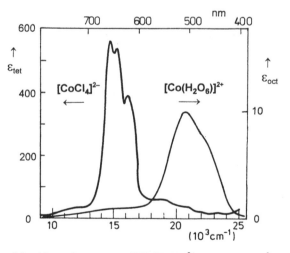

Figure 3.8 Absorption spectra of $[Co(H_2O)_6]^{2+}$ and of $[CoCl_4]^{2-}$.

A lot of experimental work has been devoted to this problem, mostly concentrating on absorption spectra and electrical conductivity in order to identify specific complex ions present [23, 24]. The spectra of these two species are shown in Figure 3.8 and their dependence on the kind of solvent will be discussed in Chapter 5.

A strong absorption band at approximately 660 nm ($\varepsilon = 500\ M^{-1}\,cm^{-1}$) in the spectrum of the tetrahedral species results in a deep blue colour. This colour does not provide evidence for the presence of just this one complex ion, $[CoCl_4]^{2-}$. There may also occur other forms, for instance in aqueous solutions $[CoCl_2(H_2O)_2]$ or $[CoCl_3(H_2O)]^-$, although their spectra, and especially the intensity of bands, are different.

In the system under discussion, the intensity of absorption bands is an important parameter, whose value provides evidence that a blue colour characterises tetrahedral species. Owing to the lack of a centre of symmetry in these ions, the d-d electronic transition is no longer forbidden. The molar absorption coefficients in the Co(II) aqua-complex are approximately 5; in $CoCl_4^{2-}$, about 400–$500\ M^{-1}\cdot cm^{-1}$.

Let us note, however, that a blue colour cannot be automatically taken as a proof of the presence of tetrahedral forms. One exception cited in the literature [25] is solid anhydrous $CoCl_2$, which is deep blue despite having the $CdCl_2$ structure with the Co^{2+} ion in an octahedral environment of six chloride ions. A complementary example is provided by the dipivaloylmethanate anion, which forms a pink tetrahedral complex with Co^{2+}. Furthermore, we have shown [26] that if the spectrum of the hypothetical ion $[CoCl_6]^{4-}$ is simulated (with absorbance values calculated for a concentration of $1\ mol\,dm^{-3}$ and molar absorption coefficients the same as for $[Co(H_2O)_6]^{2+}$) the calculated chromaticity coordinates correspond exactly with the blue colour!

The explanation seems fairly straightforward. What is crucial is the strength of the field created by the ligand or, for practical purposes, its position in the spectrochemical ligand series. As a result of a large f value of the ligand in the MX_6^{n-} (O_h) complex, even though a change to T_d symmetry in the MX_4^{n-} complex shifts the spectrum bathochromically, the shift may not necessarily be as substantial as to lead to strong absorption in the 600–650 nm range, which would cause the appearance of a blue colour.

A change of the coordination number and an equilibrium of the octahedral–square-planar type are characteristic of the d^8 electronic configuration metal ions, for instance Ni(II) or Pt(II). In the case of Ni(II), this will be partially discussed when we consider the effect of solvents (solvatochromism).

Figure 3.9 presents the spectra of several Ni(II) complexes with octahedral and with square-planar structures [25]. The colour of the aqua-complex is green, that of the $[Ni(en)_3]^{2+}$ ion is purple, while that of $[Ni(en)_2)]^{2+}$ is orange. Jørgensen [27] found that in the case of the bluish-purple complex $[Ni(trien)(H_2O)_2]^{2+}$ {where the ligand trien = $H_2N(CH_2)_2NH(CH_2)_2NH(CH_2)_2NH_2$}, addition of a large amount of $NaClO_4$ resulted in the appearance of an orange hue:

$$[Ni(trien)(H_2O)_2]^{2+} \leftrightarrows [Ni(trien)]^{2+} + 2H_2O$$

bluish-purple   orange

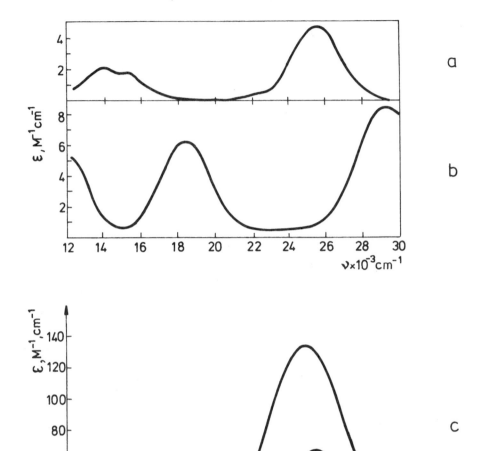

Figure 3.9 Absorption spectra of nickel(II) complexes: (a) $[Ni(H_2O)_6]^{2+}$, (b) $[Ni(en)_3]^{2+}$, (c) $[Ni(en)_2]^{2+}$ (top curve – actual spectrum; lower curves – resolved components).

The same effect was obtained by Sone and Kato [28] when they studied a mixed Ni(II) complex with propylenediamine, $NH_2CH_2CH(CH_3)NH_2$, and water. The blue or bluish purple crystals of the complex lost two water molecules not only when slightly heated but even when dried in a desiccator for a longer period of time. Orange, highly hygroscopic, crystals were obtained, which changed back into a bluish purple compound when exposed to moisture and slight heating.

Recently Bartecki and colleagues [29] studied the system $[Ni(en)_3]^{2+}$-acetone. They found that after a certain time the chelate $[Ni(en)_2]^{2+}$ appeared and they presented its characteristic absorption spectrum.

It can be concluded that in Ni(II) compounds, if the geometry changes from octahedral to square planar, the colour changes in the opposite direction than in the case of an octahedral → tetrahedral change. In crystal field theory, it is assumed that, roughly, $10Dq(sq.pl.) \cong 1.3 \times 10Dq(oct)$. Thus, it can be expected that the spectrum will undergo a hypsochromic shift. In the spectra of square-planar Ni(II) complexes, there is only one band in the approximately 400–600 nm range and a transmittance maximum at about 550–650 nm.

Generally, it is assumed that octahedral → square-planar changes are facilitated by:
a) an increase of the ligand field strength in the equatorial plane of the complex,
b) an increase of the difference in field strength between equatorial and axial ligands,
c) steric hindrances which may occur between such ligands, for instance by introducing as equatorial ligands molecules with complex long chains. For this reason, for example, trien is a more effective ligand than en in forming square-planar complexes [28].

3.2 COLOUR OF TRANSITION METAL COMPOUNDS DUE TO CHARGE-TRANSFER (CT) TRANSITIONS

3.2.1 General

As was concluded in the preceding chapter, transition metal compounds to a large extent owe their colour to ligand field transitions in the partially filled d subshell. But there are numerous examples of d^0 or d^{10} electronic configurations which show quite distinct coloration, with a very intense and at the same time characteristic colour in some cases. The purple solutions of $KMnO_4$, which – as is well known – are used for analytical purposes in redox reactions, are a good example here. $KMnO_4$ is an example of a compound whose colour is the result of charge transfer (CT) from O^{2-} to the empty d orbital of the metal ion.

There are two main types of CT transitions: from the ligand to the metal (CTLM or LMCT) and from the metal to the ligand (CTML or MLCT). There are also other possible transitions, as for instance transitions between ions of two different metals or the same metal in different oxidation states (MMCT or IVCT). The situation that will be discussed below (in connection with the colour of minerals) involves for example such transitions as $Ti^{4+} - Fe^{2+}$ and $Fe^{3+} - Fe^{2+}$. Another type of transition is ITCT (or LLCT), which occurs between two different ligands through the central metal ion. The importance of CT transitions derives from the fact that their intensity is much greater than the intensity of ligand field transitions. Table 3.11 compares different types of transitions.

Thus, from a technological point of view, it is more useful to use complexes whose absorption spectra are much more intense, as a specific colour of required purity may be obtained by using a much smaller amount of the substance. It is even more

THE COLOUR OF METAL COMPOUNDS

Table 3.11 Intensity of various electronic transitions in complexes of d-elements

Transition type	Approximate oscillator strength, f	Approximate value of ε, $M^{-1}cm^{-1}$
d-d		
O_h spin–allowed	10^{-5}	10–100
spin–forbidden	10^{-7}	0.01–0.1
T_d spin–allowed	10^{-3}	10^2
Charge–transfer, CT (LMCT, MLCT)	10^{-1}	10^4
In the ligand		10^2-10^5

important to use CT transitions in qualitative and quantitative analysis. Of course what is of significance is in what range CT bands occur. In transition metal complex compounds, such bands are usually located in the ultraviolet range, but particularly in complexes with organic ligands, there may be bands in the visible range.

CT transitions may be interpreted within molecular orbital theory. One of the first interpretations of such transitions in hexahalides of $4d$ and $5d$-electronic elements was given by Jørgensen [30]. The simplified diagram of molecular orbitals in Figure 3.10 presents the possibility of four CTLM transitions involving the t_{2g} and e_g orbitals of the metal and the σ and π orbitals of the ligand:

$$\nu_1 \pi L - M t_{2g}(\pi^*)$$
$$\nu_2 \pi L - M e_g(\sigma^*)$$

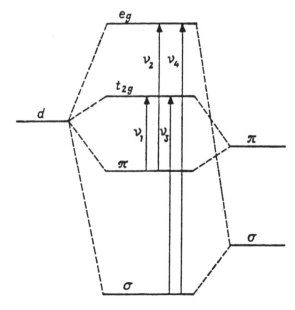

Figure 3.10 Simplified molecular orbital diagram for hexahalogenides of $4d$- and $5d$-elements [30].

$$v_3 \sigma L - M t_{2g}(\pi^*)$$

$$v_4 \sigma L - M e_g(\sigma^*),$$

where the following inequality obtains:

$$\nu_1 < \nu_2 < \nu_3 < \nu_4.$$

It has been found that generally transition energy increases in the subgroup from $4d$-electronic to $5d$-electronic elements by approximately 8000–9000 cm^{-1} and is greater for higher oxidation states of the metal ion. An important practical observation reveals that transition energy decreases in the ligand series: Cl$^-$ > Br$^-$ > I$^-$, as the ionisation potential of the halogen decreases. Thus, it can be expected that bromides and iodides, and sometimes chlorides too, can be coloured, while fluorides should be colourless. The absorption band occurring in the visible range corresponds to the ν_1 transition, with the possibility of splitting and additional bands occurring in the case of bromides and iodides due to the substantial spin-orbit coupling constant in those halogens.

Tables 3.12–3.14 give the colours of the solid halides and oxohalides of some transition elements. As far as CTML transitions are concerned, it must be said that

Table 3.12 Colour of halides of some d^0 – transition elements

Ti	TiF$_4$	TiCl$_4$	TiBr$_4$	TiI$_4$
	white	colourless	orange	dark brown
Nb	NbF$_5$	NbCl$_5$	NbBr$_5$	NbI$_5$
	white	yellow	orange	brass coloured
Ta	TaF$_5$	TaCl$_5$	TaBr$_5$	TaI$_5$
	white	white	pale yellow	black
W	WF$_6$	WCl$_6$	WBr$_6$	–
	colourless	dark blue	dark blue	

Table 3.13 Colour of some halides and oxohalides of transition elements in relation to oxidation state

VF$_5$	VF$_4$	VF$_3$	VF$_2$	
colourless	green	yellow green	blue	
d^0	d^1	d^2	d^3	
CrF$_6$	CrF$_5$	CrF$_4$	CrF$_3$	CrF$_2$
yellow	red	green	green	green
d^0	d^1	d^2	d^3	d^4
VOF$_3$	VOCl$_3$	VOBr$_3$		
yellow	yellow	deep red	d^0	
VO$_2$F	VO$_2$Cl			
brown	orange			
VOF$_2$	VOCl$_2$	VOBr$_2$	d^1	
yellow	green	yellow brown		
CrOF$_4$				
red			d^0	
CrO$_2$F$_2$	CrO$_2$Cl$_2$	CrO$_2$Br$_2$		
violet	red	red		

Table 3.14 Colour of some Mn halides and oxohalides

Oxidation state	Formula	Colour
+4	MnF_4	blue
+3	MnF_3	red purple
+2	MnF_2	pale pink
+2	$MnCl_2$	pink
+2	$MnBr_2$	rose
+2	MnI_2	pink
+7	MnO_3F	dark green

Table 3.15 CTML band positions (cm^{-1}) in some complexes with pyridine-N-oxide and with picoline oxide

Complex	Metal electronic configuration	$t_{2g} \rightarrow \pi^*$
$[Co(pyNO)_6]^{2+}$	$t_{2g}^5 e_g^2$	24 450
$[Co(4\text{-}NO_2pyNO)_6]^{2+}$	$t_{2g}^5 e_g^2$	21 100
$[Co(2\text{-}MepyNO)_6]^{2+}$	$t_{2g}^5 e_g^2$	26 600
$Co(picO)_2 \cdot 2H_2O$	$t_{2g}^5 e_g^2$	23 800
$Ni(picO)_2 \cdot 2H_2O$	$t_{2g}^6 e_g^2$	24 400

such transitions occur if there are low orbitals in the ligands to which an electron of the central metal ion can be excited. Such ligands include chiefly some nitrogen containing ligands, such as pyridine, 2, 2′-bipyridyl, or 1,10-phenanthroline, particularly if they occur in complexes with easily oxidisable metals, such as Cu(I), Fe(II), or Ti(III). However, there are also complexes of Co(II) and Ni(II) with, for example, pyridine N-oxide, where such transitions occur in the visible range: they include both $t_{2g} \rightarrow \pi^*$ and $\sigma_g \rightarrow \pi^*$ transitions (in octahedral complexes).

Some data concerning the wavenumbers of the transitions under discussion are presented in Table 3.15. It can be presumed that all the complexes are coloured, but the information is incomplete.

It has been found that the energies of CT transitions (for the same ligands) depend on the coordination number and the geometrical structure. In complexes with a d^0-electronic configuration, it has been shown that for instance in Ti(IV) halides and TiX_6^{2-} halo-complexes the energy of the CT transition (first band) shows a strong bathochromic shift for complex ions in comparison with the simple halides. According to Lever, the energies (expressed in wavenumbers) are as follows [8]:

$TiCl_4$	$35\,600\,cm^{-1}$	$TiCl_6^{2-}$	$25\,000\,cm^{-1}$
$TiBr_4$	$29\,500\,cm^{-1}$	$TiBr_6^{2-}$	$21\,800\,cm^{-1}$
TiI_4	$19\,600\,cm^{-1}$	TiI_6^{2-}	$12\,100$ and $14\,300\,cm^{-1}$

This is explained by the fact that for instance in liquid $TiCl_4$ the covalence of the bond is much greater than in the $TiCl_6^{2-}$ complex ion. The effect of the crystal field in T_d and O_h symmetries is that the stabilisation of the acceptor orbital of the Ti(IV) ion is equal to $-4Dq$ in the former case and $-6Dq$ in the latter case. When this energy is considered it may be shown that the shift in the halide series corresponds with a change of so-called optical electronegativity by approximately 0.3 units.

Optical electronegativity is an important notion which enables an estimation of CT transition energies. The value, introduced by Jørgensen [7], is connected with the change of the CTLM transition energy in the F, Cl, Br, I series and is successively 28 000, 6000, and 10 000 cm^{-1} lower. These shifts are proportional to the differences in electronegativity according to Pauling's scale, if 30 000 cm^{-1} is regarded as a value equivalent to one unit of that scale.

Thus, optical electronegativity can be defined by the following equation:

$$\nu_p = (x_L - x_M) \times 30000 \text{ cm}^{-1},$$

where x_L and x_M are the optical electronegativities of the ligand and of the metal, and ν_p is the wave number of the first CT transition.

3.2.2 Colour of transition metal d^0 oxoanions and oxocations

In our discussion of the spectrochemical series of ligands we mentioned some difficulties with the positioning of the O^{2-} ion as a ligand. However it seems that in the case of transition metal oxoanions and oxocations with the d^0 electronic configuration the situation is quite clear. In other words, when considering MO_x^{n-} and MO_x^{n+} systems {listed in Table 3.16 (oxoanions) and Table 3.17 (oxocations)}, we

Table 3.16 Examples of oxoanions and isopolyanions of 3d elements

VO_4^{3-}	CrO_4^{2-}	MnO_4^-
VO_3^-	$Cr_2O_7^{2-}$	MnO_4^{2-}
$V_2O_7^{4-}$	$Cr_3O_{10}^{2-}$	MnO_4^{3-}
	MoO_4^{2-}	TcO_4^-
	$Mo_7O_{24}^{6-}$	
	WO_4^{2-}	ReO_4^-
	$W_2O_8^{4-}$	

Table 3.17 Mononuclear oxocationic units of formula MO_x^{n+}

		n	x = 1	n	x = 2	n	x = 3
IV	d^0	2	$TiO^{2+}, ZrO^{2+}, HfO^{2+}, (ThO^{2+})$				
V	d^0	3	$VO^{3+}, NbO^{3+}, TaO^{3+}, (PaO^{3+})$	1	VO_2^+, NbO_2^+, TaO_2^+		
	d^1	2	VO^{2+}				
VI	d^0	4	$(CrO^{4+}?), MoO^{4+}, WO^{4+}$	2	$CrO_2^{2+}, MoO_2^{2+}, WO_2^{2+}$		
	d^1	3	$CrO^{3+}, MoO^{3+}, WO^{3+}$				
VII	d^0	5	ReO^{5+}	3	ReO_2^{3+}	1	$MnO_3^+, TcO_3^+, ReO_3^+$
	d^1	4	ReO^{4+}	2	ReO_2^{2+}, MnO_2^{2+}		
	d^2	3	TcO^{3+}, ReO^{3+}	1	ReO_2^+		
VIII	d^0					2	OsO_3^{2+}
	d^1	5	OsO^{5+}				
	d^2	4	RuO^{4+}, OsO^{4+}	2	RuO_2^{2+}, OsO_2^{2+}		
	d^4	2	RuO^{2+}				

Table 3.18 Observed transition energies in some oxoanions and their assignments by various authors

	MnO_4^-		CrO_4^{2-}		VO_4^{3-}	
	I	II	I	II	I	II
Transition energies (kK)	18	32	26	36	37	(45)
Assignments[a]:						
WH	$t_1 \to 3t_2$	$2t_2 \to 3t_2$	$t_1 \to 3t_2$	$2t_2 \to 3t_2$		
BL	$t_1 \to 2e$	$t_1 \to 3t_2$	$t_1 \to 2e$	$t_1 \to 3t_2$		
VG	$t_1 \to 2e$	$2t_2 \to 2e$				
		$t_1 \to 3t_2$				

[a] WH = Wolfsberg and Helmholz (ref. [31]); BL = Ballhausen and Liehr (ref. [32]); VG = Viste and Gray (ref. [33]).

assume these entities to be a chromophoric group responsible for the colours of the corresponding compounds. A considerable amount is known about these oxospecies, but it should not be assumed that the elucidation of the electronic transitions which form the basis of the colours observed is always firmly established. The most thoroughly examined species of this type are the tetrahedral oxoanions MO_4^{n-}, particularly MnO_4^- and CrO_4^{2-}. On the basis of molecular orbital theory, two bands were predicted [31, 32, 33, 34] but differently assigned by various authors. Generally speaking these bands are due to $p\pi(O) \to d\pi(M)$ electronic transitions. Approximate values of transition energies and MO assignments are given in Table 3.18 [31, 32, 33]. Attempts have been made to find a simple function to express the dependence of these energies on properties of the central metal. Many years ago Symons [35] recalled an earlier suggestion by Teltow [36] that the energy of the long-wave absorption band (called the first band) of tetraoxoanions was a function of the ionic radius of the metal. Bartecki and coworkers have, as a result of their extensive studies on oxocations and oxoanions, shown that this function is of more general applicability. Firstly, it is valid not only for the first long-wave band but also for the second short-wave band. Secondly, the dependence was found to apply also to MO_2^{n+} oxocations. Basic arguments to justify the role of ionic radii were given at the XIII International Conference on Coordination Chemistry in 1970 [37]. Plots of the two absorption maxima, both for oxoanions and for oxocations, versus ionic radii are depicted in Figure 3.11.

The straight line functions may be represented to a good approximation by the following equations linking absorption frequency σ (in kK) with ionic radius r (in Å):

$$\text{(I)} \quad \sigma = 110r - 38.1 \quad\quad \text{(II)} \quad \sigma = 162r - 58.9$$
$$\text{(III)} \quad \sigma = 158r - 53.5 \quad\quad \text{(IV)} \quad \sigma = 97r - 11.8$$

Table 3.19 presents the calculated and experimental values for the transition energies in oxoanions and oxocations.

Our main target now is to see how far the band positions of these entities, and of the corresponding compounds, are in line with the resulting colour. To be precise, it is not only band position which is important but also band contour, in the visible and near infrared. However, absorption maxima (or absorption minima – as has already been discussed earlier in this book) provide important hints. This problem is very much neglected with colour being said to be to all intents and purposes

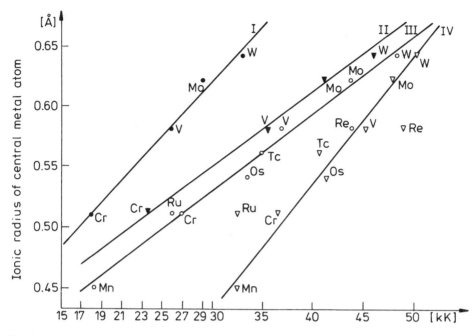

Figure 3.11 Plots of the energies (in kK) of the two absorption maxima of transition metal oxocations (I, II) and oxoanions (III, IV) against ionic radii of the central metal atoms.

Table 3.19 Calculated and experimental values of transition energies (kK) in oxoanions, tetroxides, and oxocations

	R_z	Band I		Band II		II–I	
		calc.	obs.	calc.	obs.	calc.	obs.
OXOANIONS							
MnO_4^-	0.45	18.50	18.32	31.85	32.21	13.35	13.89
CrO_4^{2-}	0.51	27.08	26.81	37.67	36.63	10.59	9.82
TcO_4^-	0.56	34.98	34.90	42.52	40.70	7.54	5.80
VO_4^{3-}	0.58	38.14	36.90	44.46	45.30	6.32	8.40
ReO_4^-	0.58	38.14	44.00	44.46	49.00	6.32	5.00
MoO_4^{2-}	0.62	44.46	43.90	48.34	48.00	3.88	4.10
WO_4^{2-}	0.64	47.62	48.50[a]	50.28	50.30[b]	2.66	1.80
OXIDES							
RuO_4	0.51	27.08	26.00	37.67	32.20	10.59	6.20
OsO_4	0.54	31.82	33.50	40.58	41.50	8.76	8.00
OXOCATIONS							
CrO_2^{2+}	0.51	18.00	18.87	23.72	24.00	5.72	5.13
VO_2^{2+}	0.58	25.70	26.00	35.06	36.00	9.36	10.00
MoO_2^{2+}	0.62	30.10	29.90	41.54	41.60	11.44	11.70
WO_2^{2+}	0.64	32.30	32.40	44.78	46.50	12.48	14.10

[a] Ref. [42]; [b] Ref. [35].

determined by band positions. For example A.F. Williams in his book [38] simply links the changes in colour from colourless VO_4^{3-} to yellow CrO_4^{2-} to violet MnO_4^- with the respective absorption maxima at 36 900, 26 800, and 18 000 cm^{-1}. However, the second band of the permanganate ion, although its absorption maximum is significantly distant at 32 000 cm^{-1}, does influence the colour of this anion through the strong absorption by its long-wave tail. Additive mixing of the transmitted light corresponding to the two minima at around 420 nm (purple blue) and 620 nm (red) gives rise to the characteristic red purple or purple colour. Note that, despite its frequent use, for example by Newton (p. 1) and more recently (see, e.g. Tables 3.1, 4.4, 5.4, 5.5, and 8.2), including the present authors, the term 'violet' is not officially recognised (CIE; note its absence from Figure 1.11).

As in the case of permanganate, hundreds of papers have been devoted to the spectroscopic characteristics of the chromate ion, which as an isolated ion has perfect T_d symmetry. Two main absorption bands appear in the spectrum of, e.g., aqueous solutions of alkali metal chromates (Figure 3.12). The first band, at approximately 26 000 to 26 800 cm^{-1}, and the second, at approximately 36 000 cm^{-1}, could not cause the yellow colour of the solution. That the second band is not responsible for the observed colour is clear, as not even the tail of this band extends into the visible region. The chromaticity diagram shows that a yellow colour for an object corresponds to a blue complementary hue in the region of 480 nm (20 800 cm^{-1}).

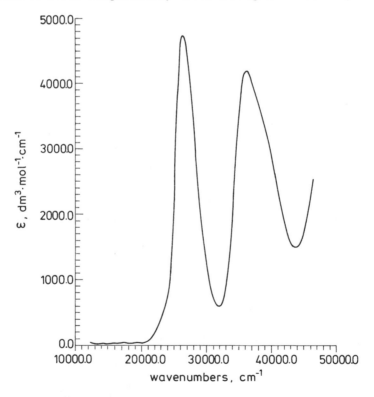

Figure 3.12 The absorption spectrum of an aqueous solution of K_2CrO_4 (1×10^{-3} mol dm^{-3}).

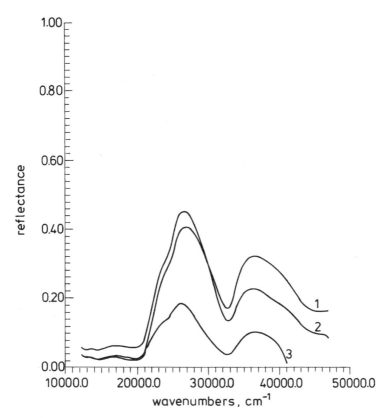

Figure 3.13 Reflectance spectra of K_2CrO_4 (5% by weight) in Li_2CO_3 (spectrum 1), BaO (spectrum 2), and MgO (spectrum 3) matrices.

In a recent paper [39] an analysis of the spectrum of an aqueous solution of potassium chromate by curve resolution using a digital filtration method was reported. An additional band was found at about $16\,500\,cm^{-1}$, whose tail probably reaches a minimum at about $20\,000\,cm^{-1}$ (480 nm), giving rise to the observed yellow colour.

Chromate salts have also been studied in the solid state. In the paper cited above [39], spectroscopic and colorimetric investigations involving various matrices were reported. The site symmetry of the chromate ion in the solid state is no longer precisely tetrahedral and indeed depends on the matrix. The band positions are not significantly affected in MgO, Li_2CO_3, or BaO matrices, but in other matrices they are more or less affected. Some reflectance spectra from this paper are reproduced in Figure 3.13.

In order to characterise the colour, CIE, CIELAB, and CIELUV chromaticity coordinates of all the systems studied were calculated, as well as the hue angle h_{ab}. The results are depicted in Figure 3.14, which shows illustrations of the CIELAB plane of colour points for mixtures containing 5% by weight of K_2CrO_4 or Ag_2CrO_4 in all matrices.

When comparing the three oxoanions mentioned, viz. VO_4^{3-}, CrO_4^{2-}, and MnO_4^{-}, and taking into account the results on colour one could reasonably conclude that

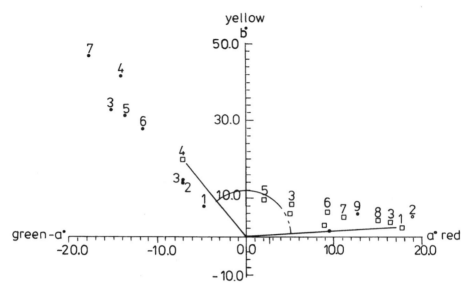

Figure 3.14 The CIELAB (1976) plane. The points correspond to 5% (by weight) mixtures of K_2CrO_4 (•) and Ag_2CrO_4 (□) in matrices of: 1. MgO; 2. BaO; 3. Li_2CO_3; 4. KBr; 5. CsBr; 6. Me_4NBr; 7. KCl; 8. CsCl; 9. Ag_2SO_4. The angles show the extreme values of the colour parameters for Ag_2CrO_4.

increasing atomic number Z of the 3d elements is in line with the shift of colours along the right-hand side of the chromaticity diagram, from yellow to purple.

We can now look at some other colour systematics for d^0-oxygen systems, for instance the homologous entities for the 3d, 4d, and 5d electronic configurations. As an example we can try to understand the change of colour in the series CrO_4^{2-}, MoO_4^{2-}, and WO_4^{2-}, or the series WO_4^{2-}, WS_4^{2-}, and WSe_4^{2-}. Still another possibility arises when comparing series of mixed species, e.g. CrO_4^{2-}, CrO_3X^-, CrO_2X_2, and $CrOX_3^+(X = F, Cl, Br)$. We shall first consider spectra of monosubstituted CrO_3X^- ions, which have been the object of many studies, particularly in organic solvents and at low temperatures (liquid nitrogen; single crystals at liquid helium temperature) [40, 41]. The electronic spectra of $K[CrO_3Cl]$ in acetone, acetonitrile, dimethylform-amide, dimethylsulphoxide, and tetrahydrofuran (abbreviated as A, AN, DMF, DMSO, and THF) were studied in detail. Figure 3.15 shows the spectrum in AN solution, Figure 3.16 the diffuse reflectance spectrum in Li_2CO_3. In most cases four absorption regions could be identified, at about 22 000, 28 000, 36 000, and 42 000 cm^{-1} (450, 355, 280, and 240 nm). In the case of aprotic solvents all electronic bands appear to have fine structure, i.e. these transitions are electronic-vibrational in character. The spectra of the fluoro- and bromo- analogues are of very similar character, though with minor differences in the ultraviolet region. There appears to be no substantial difference from the behaviour of the parent chromate ion, as others have also concluded [34]. However, the colour of the CrO_3X^- anions is orange-yellow rather than the yellow of CrO_4^{2-}.

We proceed now to the series of thioanions and mixed oxothioanions of molybdenum and tungsten. Considering the tetrathioanions of the two elements, one

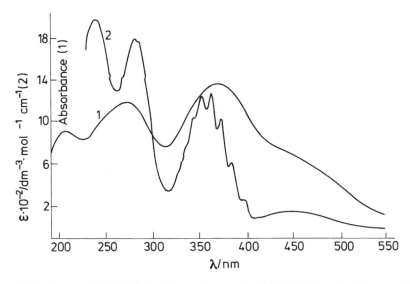

Figure 3.15 Spectra of KCrO₃Cl: 1. diffuse reflectance, in Li₂CO₃; 2. absorption, in acetonitrile.

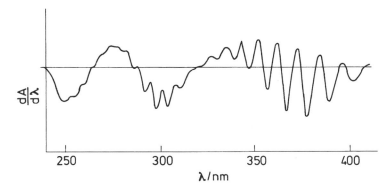

Figure 3.16 The first derivative spectrum of KCrO₃Cl in acetonitrile.

can easily predict that there should be a marked bathochromic shift of the spectra in relation to the tetraoxoanions, and simultaneously a hypsochromic shift from Mo^{VI} to W^{VI} thioanions, just as in the case of the oxoanions of these elements. These statements may be justified as follows:

(1) applying Jørgensen's formula $\sigma(LMCT) = 30\,000(X_L - X_M)$, the transition energy must be greater in oxoanions than in thioanions as X_O is greater than X_S;

(2) according to the same formula there should be a bathochromic shift in the transition energies when comparing Mo^{VI} and W^{VI} optical electronegativities.

Figure 3.17 shows the uv-visible absorption spectrum of the piperidinium salt of tetrathiomolybdate, $(C_5H_{11}NH)_2MoS_4$, in water [42]. Gaussian analysis of the

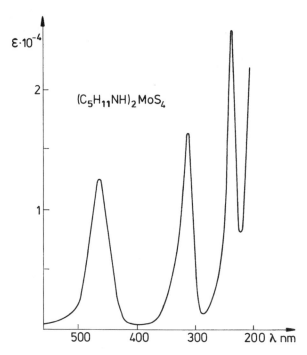

Figure 3.17   Absorption spectrum of $(C_5H_{11}NH)_2MoS_4$, in aqueous solution (10^{-4} mol dm^{-3}).

electronic spectra of $(NH_4)_2MoS_4$ and of $(NH_4)_2WS_4$ gave values for the frequencies of maximum absorption at 21 400, 31 600, and 41 400 and at 25 450, 36 000, and 46 900 cm^{-1} respectively [42]. It is interesting to note that the first two absorption bands, as obtained by Gaussian analysis, conform to the dependence of transition band energies on central atom ionic radii, giving two straight lines for vanadium, molybdenum, tungsten, and, for the lower band, rhenium (for the respective highest oxidation states) [42].

Next we consider metal oxocations, chiefly those of d^0 configuration, as a group of chromophores. As chemical entities they form two- or three-atom cations of the type $MO_x^{n+}(x = 1$ or 2). There are only a few oxocations which should be regarded as, or have been shown to be, analogues of monoatomic cations. This book is devoted to colour problems and from this standpoint it is important to search for cases in which the oxocationic group and its colour are firmly established. The best known practically, and well understood theoretically, case is the uranyl cation, UO_2^{2+}, which has an f^0 electronic configuration. Hundreds, if not thousands, of papers have been published on its chemistry, photochemistry, and spectroscopy. One of the authors of this book has long been engaged in research into electronic spectra of this cation in organic solvents [43]. As we are here mainly interested in the d-block elements we shall merely stress that the colour of uranyl compounds is almost always yellow, except in cases where there is a high covalent contribution to bonding. Thus, for example, the acetylacetonate (2,4-pentanedionate) complex is nearly red. Generally the spectra of uranyl complexes can be characterised in terms of vibrational structure

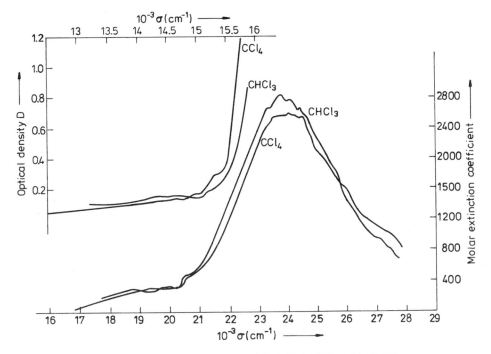

Figure 3.18 Absorption spectrum of CrO_2Cl_2 in CCl_4 and in $CHCl_3$.

of the UO_2^{2+} molecular ion, in which the 0–0 electronic transition occurs at approximately $24\,000\,cm^{-1}$ ($\sim 416\,nm$). This value corresponds to the appearance of a yellow colour.

As previously mentioned, in the group of d^0-oxocations those containing (or assumed to contain) a triatomic molecular entity of the MO_2^{n+} type have been the object of studies by Bartecki's group [37, 38, 41, 42]. CrO_2Cl_2 was one such compound intensively studied from this point of view [44]. Its absorption spectra in CCl_4 and in $CHCl_3$ are given in Figure 3.18. These spectra are clearly reminiscent of the spectrum of uranyl nitrate, $UO_2(NO_3)_2$, because in both cases the absorption maximum is in the 410–420 nm ($\sim 24\,000\,cm^{-1}$) region, again showing some traces of vibrational structure. This band exhibits a much greater molar absorption coefficient, of about 2600. The second band has its maximum at $18\,870\,cm^{-1}$. The red colour of liquid chromyl chloride, and of its solutions in the inert solvents used in these studies, corresponds reasonably with these data. Consideration of the electronic spectra of a few chromyl compounds enables one to draw conclusions about the CrO_2^{2+} entity as a distinct chromophoric group. There is a marked colour progression from yellow to orange to red on going from CrO_4^{2-} to CrO_3Cl^- to CrO_2Cl_2.

If we proceed to the analogous compounds of molybdenum, i.e. containing the oxocation MoO_2^{2+}, we can state that most of the studies have been carried out on the simple oxohalides MoO_2Cl_2 and MoO_2Br_2, and their complexes. The main band in the absorption spectra of MoO_2Cl_2 and of MoO_2Br_2 appears, as a maximum or

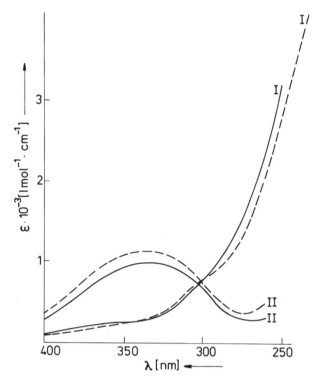

Figure 3.19   Absorption spectra of MoO_2Cl_2 (———) and of MoO_2Br_2 (– – – –) in ethanol – chloroform (1:1) mixtures. MoO_2Cl_2 concentrations are 0.002 (I) and 0.0001 (II) $mol\,dm^{-3}$; MoO_2Br_2 concentrations are 0.001 (I) and 0.0001 (II) $mol\,dm^{-3}$.

shoulder, in the range 345 to 370 nm (27 000 to 29 000 cm^{-1}; Figure 3.19). This band should be assigned to internal transitions of the MoO_2^{2+} group, hence this oxocation should exhibit chromophoric properties [45, 46]. The long-wave tail of this first band reaches its absorption minimum at about 560–570 nm and is responsible for the yellow hue of the solution. It seems that this shift in spectrum and colour relative to analogous chromyl compounds can be accommodated within the general relationship between transition energies and atomic number (for a given structure – here both oxocations have *cis*-structures and thus C_{2v} symmetry).

3.3   SOME SPECIFIC ISSUES CONNECTED WITH THE COLOUR OF TRANSITION METAL COMPOUNDS

In the previous chapter we presented some general and fundamental issues connected with the colour of transition metal compounds due to *d-d* transitions in the crystal field as well as CT transitions. As was mentioned in the introduction, until now quantitative colour measurements of transition metal compounds have not been dealt with in the literature. Quantitative colour measurements of other chemical compounds have not been discussed either, with the exception of a few papers [47, 48]. This chapter will discuss some of the work done by the team of one of the authors (A.B.) which has been devoted to the determination of chromaticity coordinates using the CIE 1931 trichromatic colorimetry method and subsequent modifications of that method.

3.3.1   Chromaticity coordinates, dominant wavelength, and optical spectra of some transition metal complex compounds

3.3.1.1   Simulation of optical spectra and chromaticity coordinates

Let us remember that while the colour of an object is connected with the absorption (or transmission) spectrum envelope, in the first approximation colour is determined by absorption and transmission areas.

Thus, by simulating absorption spectra, we can also calculate chromaticity coordinates and on that basis determine the psychophysical colour in accordance with the chromaticity diagram.

Under certain assumptions based on well-known experimental facts, one can simulate the spectrum contour of transition metal complexes with different electronic configurations and symmetries of the central ion for the assumed intensities and widths of absorption bands. The spectra of aqua-complexes, which (particularly in the case of perchlorate salts) contain $[M(H_2O)_6]^{n+}$ ions (n = 2 or 3), may serve as the basis for the simulation of the absorption spectra of $3d$-electronic compounds. Figure 3.20 presents the spectra of some $3d$-electronic compounds.

A. Simulation of CrL_6^{3+} spectra

On the basis of literature data, as well as the author's own work, the following parameters were adopted for the simulation of $[Cr(H_2O)_6]^{3+}$:

$$Dq = 1740\,\mathrm{cm^{-1}}, \qquad B = 725\,\mathrm{cm^{-1}}, \qquad \nu_1 = 17\,400\,\mathrm{cm^{-1}},$$
$$\nu_2 = 24\,587\,\mathrm{cm^{-1}}, \qquad \varepsilon_1 = 12.5\,\mathrm{M^{-1} \cdot cm^{-1}}, \quad \varepsilon_2 = 15.6\,\mathrm{M^{-1} \cdot cm^{-1}},$$
$$\Delta\nu_{1/2}^1 = 1550\,\mathrm{cm^{-1}}, \quad \Delta\nu_{1/2}^2 = 2230\,\mathrm{cm^{-1}}$$

This set of parameters (obtained through Gaussian resolution of the experimental spectrum) was taken as constant, except the value of Dq, which was varied within a broad range. For each new value of Dq, the overall contour (271 points in all) was calculated and then the chromaticity coordinates were determined using the COLOR programme [49]. The absorbance values were taken for 1 M concentration. The relevant data are presented in Table 3.20, while Table 3.21 shows the results of simulation for some ligands with known f values.

The data in Table 3.20 indicate that the colour in the chromaticity diagram changes clockwise with the decrease of Dq. The colour shifts from green yellow, to yellow, orange red, red, reddish purple, green, again green yellow, and for the smallest Dq, bluish purple.

When evaluating the hue of colour, it should be remembered that it refers to a specific concentration of the chemical compound and that the anion of the compound also has a concrete effect. These issues are discussed in the next section.

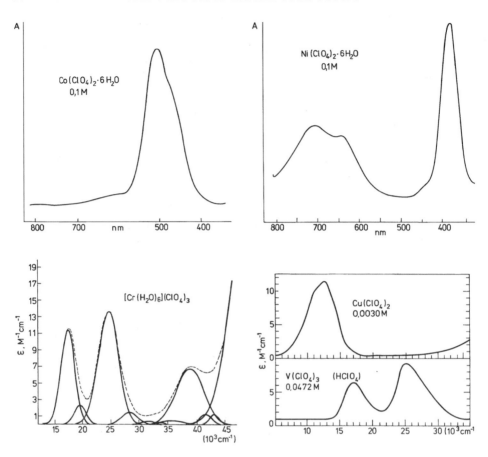

Figure 3.20 Absorption spectra of aqueous solutions of 3*d* ions (in part from [5]).

The CIE 1931 colour given in Table 3.21 clearly shows that the characterisation of the colour of the complex compounds of a given transition element with one name (as is often practised in the literature) is a remote, and possibly misleading, approximation. Of course, the assumption adopted for the simulation cannot be taken as strictly corresponding to the situation in the actual compounds of, in this case, Cr(III). Furthermore, one might expect that in some cases a significant change in the intensity of the absorption bands may distort this general picture.

The data in Table 3.22 [50] show the colours of some Cr(III) complexes, which generally confirm the above conclusions. This issue is nicely illustrated by the data concerning the colours of minerals and solid solutions containing Cr^{3+} (cf. Section 3.3.2.3 and Chapter 7). Selected simulated spectra are shown in Figure 3.21. Naturally, a decrease in Dq results in a bathochromic shift of the absorption bands, manifested in a corresponding change in colour. Both for small and for large values of Dq, there is only one band in the visible region, as the energy (wavenumber) of the first band is $10Dq$ for the d^3 configuration. Figure 3.22 shows a^* and b^* coordinates (CIE 1976) for simulated colours of several CrL_6^{3+} complexes.

Table 3.20 CIE x, y chromaticity coordinates and colour of $[CrL_6]^{3+}$ complex ions based on simulated absorption spectra as dependent on Dq values

No.	Dq cm^{-1}	Chromaticity coordinates		Colour	Dominant (or complementary*) wavelength, nm
		x	y		
1	2540	0.366	0.437	yellowish green	568
2	2440	0.411	0.503	yellow green	570
3	2340	0.454	0.521	yellow green	573
4	2240	0.508	0.486	yellow	579
5	2140	0.582	0.417	yellowish orange	591
6	2040	0.656	0.342	reddish pink	608
7	1940	0.700	0.290	red	639
8	1840	0.680	0.253	red	493*
9	1790	0.604	0.235	red purple	496*
10	1740	0.410	0.270	purplish pink	493*
11	1720	0.296	0.336	white	496
12	1690	0.146	0.506	green	505
13	1670	0.090	0.630	green	509
14	1640	0.077	0.756	green	517
15	1590	0.156	0.789	green	529
16	1540	0.257	0.726	green	544
17	1440	0.459	0.510	yellow green	573
18	1340	0.464	0.219	purplish pink	500*
19	1290	0.301	0.080	red purple	553*
20	1240	0.200	0.019	bluish purple	565*

Table 3.21 CIE colour and spectroscopic data for $[CrL_6]^{3+}$ complex ions with selected ligands (simulated absorption spectra)

L	x	y	Y	ν_1/cm^{-1}	ν_2/cm^{-1}	B/cm^{-1}	Colour
en	0.3177	0.4781	66.69	22272	29013	630	greenish yellow
NH$_3$	0.5553	0.4429	52.97	21750	28641		yellow orange
H$_2$O	0.4194	0.2729	0.052	17400	24587	727	white/purplish pink
F$^-$	0.1647	0.7870	0.130	15834	23081	765	green
CO(NH$_2$)$_2$	0.2078	0.7567	0.016	15834	22565	688	green
Cl$^-$	0.1717	0.032	0.016	13920	19311	533	purplish blue/bluish purple
Br$^-$	0.1674	0.084	0.101	13224	18099	475.6	purplish blue

B. Simulation of $Cr^{3+}-H_2O-NH_3$ spectra

Simulation of this system of mononuclear complexes is particularly interesting, as in practice mixed complexes of this kind are known. O_h symmetry was assumed even though the actual symmetry is lower. In this situation, Dq was calculated from the average environment rule, while the spectrum envelope was determined for constant absorption molar coefficients and band half-widths. Table 3.23 presents the obtained results (for 1 M concentration), which indicate the same direction of colour change as in the previous example. The hexamminechromium(III) ion is yellowish orange, mixed ions assume a succession of colours corresponding to the bathochromic shift of

Table 3.22 Colour and spectroscopic data for typical octahedral complexes of chromium(III) [50]

Complex	Colour	ν_1/cm^{-1}, $10Dq$ (λ_1 nm)	ν_2/cm^{-1} (λ_2 nm)	B, cm^{-1}
$K[Cr(H_2O)_6][SO_4]_2 \cdot 6H_2O$	Violet	17400 575	24500 408	725
$K_3[Cr(C_2O_4)_3] \cdot 3H_2O$	Reddish violet	17500 571	23900 418	620
$K_3[Cr(CNS)_6] \cdot 4H_2O$	Purple	17800 562	23800 420	570
$[Cr(NH_3)_6]Br_3$	Yellow	21500 464	28500 350	650
$[Cr(en)_3]I_3 \cdot H_2O$	Yellow	21600 463	28500 350	650
$K_3[Cr(CN)_6]$	Yellow	26700 374	32200 310	530

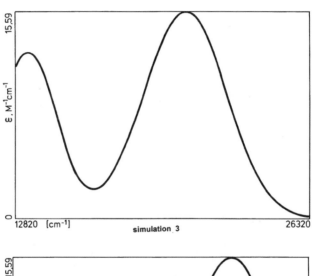

simulation 3

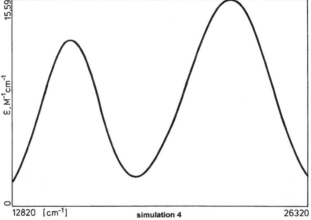

simulation 4

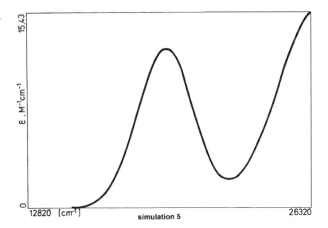

Figure 3.21 Some examples of simulated spectra for Cr^{3+}, assuming the following values for the input parameters:

Simulation:	3	4	5
$\nu_1 = 10Dq =$	13 400	15 400	19 400 cm^{-1}
$\nu_2 =$	20 600	22 600	26 600 cm^{-1}
$B =$	846	796	720 cm^{-1}
x =	0.4635	0.2560	0.6999
y =	0.2194	0.7262	0.2898
Y =	0.007765	0.12911	4.85
colour:	rose-purple	green	red

In all cases ε_1 and $\varepsilon_2 = 12.5, 15.6 \, dm^3 \, mol^{-1} \cdot cm^{-1}$; $\Delta\nu_1$ and $\Delta\nu_2 = 1550, 2230 \, cm^{-1}$; $[Cr^{3+}] = 1.0 \, mol \, dm^{-3}$; path length $= 1$ cm ($\Delta\nu$ signifies band width at half height).

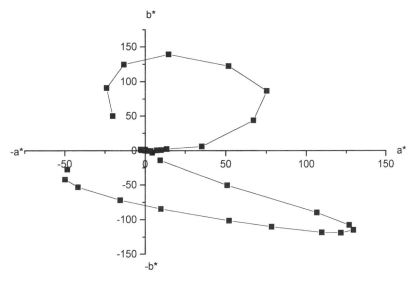

Figure 3.22 The CIELAB plane for simulated colours of CrL_6^{3+} complexes (A. Bartecki and T. Tłaczała, unpublished results presented in the lecture entitled 'Colour in Chemistry of Transition Elements' at the First Conference on Colour in Nature, Science and Technology, Wrocław, 1998).

Table 3.23 Colours and chromaticity coordinates for complexes in the $Cr^{3+}-NH_3-H_2O$ series

	$Dq(cm^{-1})$	$B(cm^{-1})$	x	y	Y	CIE colour
$[Cr(NH_3)_6]^{3+}$	2175	644	0.5553	0.4429	52.97	yellow orange
$[Cr(NH_3)_5(H_2O)]^{3+}$	2102	662	0.6110	0.3882	35.52	orange/orange red
$[Cr(NH_3)_4(H_2O)_2]^{3+}$	2030	675	0.6616	0.3882	17.13	red
$[Cr(NH_3)_3(H_2O)_3]^{3+}$	1958	688	0.6982	0.2873	4.67	red
$[Cr(NH_3)_2(H_2O)_4]^{3+}$	1885	701	0.7056	0.2716	1.66	red
$[Cr(NH_3)(H_2O)_5]^{3+}$	1813	714	0.6522	0.2440	0.26	red
$[Cr(H_2O)_6]^{3+}$	1740	727	0.4194	0.2729	0.05	rose purple

the absorption spectrum and the transmission of increasingly long wave-length radiation. Let us note that the colour of the $[Cr(H_2O)_6]^{3+}$ hexaquaion (for the assumed concentration) is purplish pink, i.e. in principle the same as the colour cited in the literature [50] e.g. for $K[Cr(H_2O)_6](SO_4)_2 \cdot 6H_2O$ (usually described as purple). The relevant data are as follows: $\nu_1 = 17\,000\,cm^{-1}$, $\nu_2 = 24\,500\,cm^{-1}$, $B = 727\,cm^{-1}$. According to the same source, for the yellow compound $[Cr(NH_3)_6]Br_3$, $\nu_1 = 21\,550\,cm^{-1}$, $\nu_2 = 28\,500\,cm^{-1}$, $B = 650\,cm^{-1}$, so the values are almost the same as those in Table 3.23 (item 1). The colour is also nearly the same.

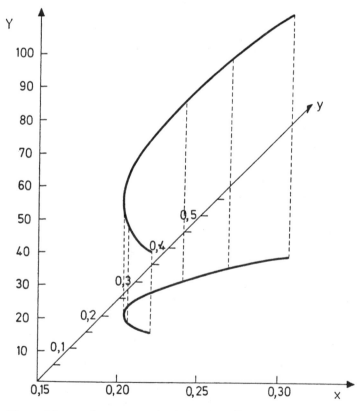

Figure 3.23 Chromaticity coordinates x, y, Y for the system $Cr^{3+}-H_2O-NH_3$ from simulated spectra.

The chromaticity coordinates in the x, y, Y system are presented in Figure 3.23. It is easy to establish that the different Dq values for the $Cr^{3+}-H_2O-NH_3$ system, introduced in Table 3.20, result in almost identical chromaticity coordinates and colour.

This kind of simulation does not enable the prediction of the colour of geometric *cis-trans* isomers. This is of course due to the fact that the average environment rule, on which the estimation of Dq is based, does not distinguish between these two situations. As in reality such isomers differ greatly in colour, e.g. *trans*-$[Co(NH_3)_4Cl_2]Cl$ is green, while the *cis* isomer is violet, changes of the tetragonal deformation parameters, Ds and Dt, or the AOM parameters should be used to explain this difference. Such attempts have not been undertaken yet.

C. Simulated spectra of the $Cr^{3+}-H_2O-Cl^-$ system

The simulation of this system leads to the observation that the introduction of one Cl^- ion to the coordination sphere of the complex causes a marked change of colour, in accordance with the position of this ion in the spectrochemical series. The $[Cr(H_2O)_5Cl]^{2+}$ .ion is green, in contrast to the purplish pink colour of the Cr(III) hexaquaion established above. Experimental data [50] confirm that the gradual introduction of Cl^- ions causes a clear bathochromic shift of the electronic spectrum. For the $[CrCl_6]^{3-}$ ion we obtain $\nu_1 = 13\,920$, $\nu_2 = 19\,310\,cm^{-1}$ from simulation experiments, and the chromaticity coordinates determine the colour as purplish blue (Table 3.24). In molten LiCl–KCl eutectic, $\nu_1 = 12\,500$, $\nu_2 = 18\,500\,cm^{-1}$ [51].

D. Simulated spectra of the $Ni^{2+}-H_2O-NH_3$ system

The data are given in Table 3.25.

As can be seen, the colour of all complexes is located in the blue, blue green or bluish green range, which is in agreement with the hypsochromic shift of the spectrum. As in the case of the previous system, the same direction of colour changes with changes of Dq can be observed. The decrease of this value from ammonia to water as the ligand causes a clockwise shift of the colour in the CIE diagram. The change cannot be great as the Dq value, estimated from the average environmental rule, changes by only about $180\,cm^{-1}$. This is due to the fact that H_2O and NH_3 are

Table 3.24 Spectroscopic data, CIE chromaticity coordinates and colour based on simulation of absorption spectra for the system $Cr^{3+}-H_2O-Cl^-$ ($[Cr^{3+}] = 1M$, $d = 1\,cm$, $\varepsilon = const.$, $\Delta\nu_{1/2} = const.$, B calculated as a mean value)

Complex	ν_1/cm^{-1} $\varepsilon_1=12.54$	ν_2/cm^{-1} $\varepsilon_2=15.56$	B/cm^{-1}	x	y	CIE colour
$[Cr(H_2O)_6]Cl_3$	17400	24587	726.8	0.4194	0.2729	orange pink
$[Cr(H_2O)_5Cl]Cl_2$	16820	23710	694.6	0.2820	0.5267	green
$[Cr(H_2O)_4Cl_2]Cl$	16240	22831	662.4	0.2063	0.7333	green
$[Cr(H_2O)_3Cl_3]$	15660	21592	630.2	0.2280	0.3337	blue green
$[Cr(H_2O)_2Cl_4]^-$	15080	21073	598.0	0.1761	0.0104	bluish purple
$[Cr(H_2O)Cl_5]^{2-}$	14500	20193	565.8	0.1729	0.0047	purplish blue
$[CrCl_6]^{3-}$	13920	19311	533.6	0.1717	0.0052	purplish blue

Table 3.25 CIE and CIELAB chromaticity coordinates and CIE colour for the system $Ni^{2+}-H_2O-NH_3$ (simulated spectra)

Complex ion	Dq [cm^{-1}]	x	y	Y	L^*	a^*	b^*	CIE colour
$[Ni(H_2O)_6]^{2+}$	930	0.2309	0.3845	72.70	104.31	−63.84	5.02	bluish green
$[Ni(H_2O)_5(NH_3)]^{2+}$	858	0.1992	0.3414	63.65	99.79	−64.56	−12.57	blue green
$[Ni(H_2O)_4(NH_3)_2]^{2+}$	987	0.1754	0.2963	53.60	94.23	−59.30	−29.04	blue green
$[Ni(H_2O)_3(NH_3)_3]^{2+}$	1015	0.1617	0.2530	43.99	88.22	−47.11	−43.43	blue green/ greenish blue
$[Ni(H_2O)_2(NH_3)_4]^{2+}$	1044	0.1583	0.2085	34.55	81.71	−25.37	−57.44	greenish blue
$[Ni(H_2O)(NH_3)_5]^{2+}$	1074	0.1680	0.1714	27.50	75.44	3.35	−68.14	blue
$[Ni(NH_3)_6]^{2+}$	1101	0.1892	0.1458	23.10	71.18	33.56	−75.09	blue

Table 3.26 ΔE_{ab}^* values for the system $Ni^{2+}-NH_3-H_2O$, (based on simulated spectra)

Consecutive ions		ΔE_{ab}^*
$[Ni(NH_3)_6]^{2+}$	$- [Ni(NH_3)_5(H_2O)]^{2+}$	8.1806
$[Ni(NH_3)_5(H_2O)]^{2+}$	$- [Ni(NH_3)_4(H_2O)_2]^{2+}$	15.6979
$[Ni(NH_3)_4(H_2O)_2]^{2+}$	$- [Ni(NH_3)_3(H_2O)_3]^{2+}$	79.4019
$[Ni(NH_3)_3(H_2O)_3]^{2+}$	$- [Ni(NH_3)_2(H_2O)_4]^{2+}$	45.8416
$[Ni(NH_3)_2(H_2O_4]^{2+}$	$- [Ni(NH_3)(H_2O)_5]^{2+}$	47.9111
$[Ni(NH_3)(H_2O)_5]^{2+}$	$- [Ni(H_2O)_6]^{2+}$	40.7075

relatively close to each other in the spectrochemical series, and the f value for NH_3 is equal to 1.25 as compared with 1.00 for water.

When changes of chromaticity coordinates are considered both in the CIE system and in the CIELAB system, it is found that the y coordinate regularly decreases with every next introduction of an NH_3 molecule, as does the luminance value Y. In the CIELAB system, there is a fair degree of regularity for L^* and for the b^* coordinate.

In Chapter 1 it was stressed that fine distinctions between colours should be determined using all three chromaticity coordinates. Equation (1.22), which defines ΔE_{ab}^*, is often used to determine differences in chromaticity, even though this issue is the subject of further research due to its particular practical significance for the *faithful* reproduction of the colours of objects [52].

The values of ΔE_{ab}^* are given in Table 3.26.

It can be seen in the table that the greatest change of ΔE_{ab}^*, characterising the difference in chromaticity, occurs between the diaquatetrammine complex and the triaquatriammine complex. At this stage it remains unclear how to interpret this fact. A hypothesis could be proposed that it is due to a dramatic change in the symmetry and geometry of the latter ion. It can also be seen that the introduction of one water molecule leads to only small changes of chromaticity in relation to the hexa-ammine complex, while the reverse exchange leads to a much larger ΔE_{ab}^* value.

Experimental results have also been obtained for the $Ni^{2+}-H_2O-NH_3$ system. Bartecki and Staszak [53] developed a method which enabled the calculation of

spectrum envelopes for all consecutive mononuclear complexes. The spectra are
shown in Figure 3.24 and the CIE and CIELAB coordinates are given in Table 3.27.

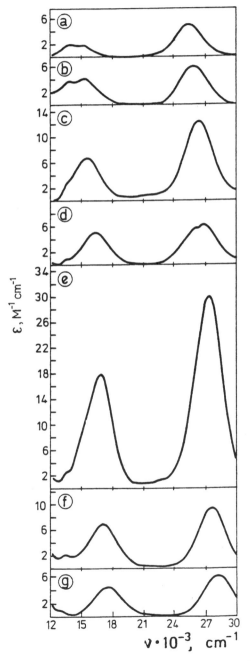

Figure 3.24   Absorption spectra of consecutive complexes in the system $Ni^{2+}-H_2O-NH_3$ [53].

Table 3.27 CIE and CIELAB chromaticity coordinates and CIE colour for the system $Ni^{2+}-NH_3-H_2O$ (experimental spectra)

Complex ion	Dq cm^{-1}	B cm^{-1}	x	y	Y	L^*	a^*	b^*	CIE colour
$[Ni(H_2O)_6]^{2+}$	930	841	0.2036	0.4121	59.67	97.66	−82.49	8.48	bluish green
$[Ni(H_2O)_5(NH_3)]^{2+}$	939	843	0.1332	0.2989	34.72	81.53	−78.52	−28.67	blue green
$[Ni(H_2O)_4(NH_3)_2]^{2+}$	981	837	0.1165	0.1952	19.13	66.84	−41.38	−55.27	bluish purple
$[Ni(H_2O)_3(NH_3)_3]^{2+}$	1175	833	0.1273	0.1359	16.35	63.43	−1.36	−77.37	blue
$[Ni(H_2O)_2(NH_3)_4]^{2+}$	1077	839	0.1379	0.0541	4.87	42.38	71.17	−101.88	purplish blue
$[Ni(H_2O)(NH_3)_5]^{2+}$	1095	833	0.1429	0.051	5.74	44.75	83.68	−111.07	blue

Both from the figures and from the tables it can be seen that there are some differences between the data. It should be emphasised that intensity changes were neglected for the simulated spectra. As can be seen the bands of the complex NiA_3B_3, where A and B are the appropriate ligands, are the most intense.

E. Simulation of spectra with variable half-width and variable intensity

We have said several times before that the colour of a compound is ultimately determined by the whole spectrum contour in the range 360–780 nm. In the discussion so far, our attention has been mostly devoted to the energies of electronic transitions and their changes resulting in changes of colour. Both half-width and absorption band intensity may also affect colour to a certain extent. Many simulations were carried out to examine the effect of those factors and the results are given in Table 3.28.

Keeping the wavenumber for one band constant (an absorption spectrum similar to the spectra of Cu(II) compounds), the two other parameters were varied. As can be seen, the most characteristic change occurs in the luminance. The Y value at constant

Table 3.28 CIE chromaticity coordinates and colour for a single absorption band (at 15000 cm^{-1}) of various intensities and band half-width (simulated spectra)

No.	Band half-width (cm^{-1})	Intensity (relative)	x	y	Y	CIE colour
1	1000	1	0.2269	0.3182	76.15	blue green/white
2	2000	1	0.1863	0.2402	44.86	greenish blue/blue
3	3000	1	0.1858	0.1830	27.25	blue
4	4000	1	0.1970	0.1723	19.42	purplish blue/blue
5	5000	1	0.2116	0.1846	15.73	purplish blue
6	10000	1	0.2698	0.2681	11.25	white
7	1000	10	0.1544	0.2675	50.33	greenish blue/blue
8	2000	10	0.1408	0.0626	7.69	purplish blue
9	3000	10	0.1600	0.0167	0.85	purplish blue/bluish purple
10	5000	10	0.1698	0.0067	0.00	bluish purple
11	10000	10	0.1686	0.0105	0.00	bluish purple
12	1000	100	0.1314	0.1980	32.03	greenish blue
13	3000	100	0.1712	0.0055	0.02	purplish blue/bluish purple
14	5000	100	0.1758	0.0200	0.00	purplish blue/bluish purple

intensity systematically decreases, and the greater the intensity, the greater the decrease. This result is easy to predict as an increase of half-width is accompanied by an increase of the electronic transition oscillator strength (which may be practically estimated on the basis of the area under the given absorption band). Thus, absorbance increases substantially, while transmittance, and hence luminance, drops significantly.

The examples cited so far (and some to be discussed in the remainder of this book) show that the simulation of absorption spectra and the prediction of the colour of complex transition metal compounds based on such simulations provide good results in a large number of cases, despite their approximate nature. It should be stressed once again that certain effects cannot be estimated in this way. First of all it is not possible to predict the colour of *trans-cis* isomers in mixed complexes or *mer-fac* isomers in MA_3B_3-type complexes if the prediction is only based on the spectro-chemical series. However, it seems possible to apply this approximation to ionisation isomers, where the coordination sphere, while variable, is uniquely determined.

The simulation of the colour of coordination ions does not make any provisions for the cation-anion interaction. It can be neglected, especially in solutions, but in the solid phase the actual colour often depends on the counterion. Thus the colour of complex salts may undergo clear changes if, for instance, the salt contains an alkaline earth metal ion instead of an alkali metal cation.

It seems that despite its simplifications simulation can be practically useful, or at least can provide preliminary or approximate data concerning the electronic spectrum envelope and the predicted colour.

3.3.1.2 *Colour and chromaticity coordinates of some 3d-electronic compounds*

The data contained in the literature concerning the colour of pigments [54] indicate that in mixture with the so-called matrix (standard white colour), so-called dichromatism[5] may occur. This problem was studied some time ago for chemical compounds [48]. That paper was probably the first to deal with the chromaticity coordinates of inorganic compounds. It was found that depending on the concentration of a substance its colour may change, which means that the chromaticity coordinates form a curve rather than a straight line.

Table 3.29 lists the CIE chromaticity coordinates for the Cr(III) nitrate, sulphate, and chloride [55], and Figure 3.25 presents the relevant curves for these and for three further compounds. In the English-language literature such curves are known as 'chromaticity traces'.

As can be seen from the data presented, only in the case of the chloride is there a linear dependence, and in the chromaticity diagram its line is situated in the green colour area. In the case of the nitrate and the sulphate, curves are obtained in different fields of the diagram with some coordinates corresponding to white colour.

The difference between the colours of these three compounds in water solutions can be predicted on the basis of differences in their absorption spectra, which are included in Figure 3.26.

---

[5] The term **dichromatism** is also used to denote dichromatic vision as a colour perception defect – see footnote 3 in Chapter 1 (Section 1.4).

Table 3.29 CIE chromaticity coordinates of aqueous solution of Cr(III) nitrate, chloride and sulphate

Concentration (mol/dm^3)	x	y	z
$Cr(NO_3)_3$			
0.02	0.30025	0.31816	0.38159
0.04	0.28763	0.30345	0.40892
0.08	0.26847	0.27406	0.45747
0.10	0.25950	0.26023	0.48027
0.20	0.23980	0.20922	0.55098
0.30	0.25889	0.19176	0.54938
$CrCl_3$			
0.02	0.29572	0.37064	0.33364
0.04	0.27521	0.41635	0.30844
0.08	0.23232	0.53969	0.22798
0.10	0.21806	0.59787	0.18406
0.20	0.16838	0.71119	0.12043
0.30	0.15325	0.75106	0.09568
$Cr_2(SO_4)_3$			
0.02	0.28211	0.29434	0.42355
0.04	0.26279	0.26608	0.47113
0.08	0.23579	0.21635	0.54786
0.10	0.23399	0.20305	0.56296
0.20	0.29328	0.18724	0.51949
0.30	0.36306	0.19446	0.44247

It can be seen that the maxima of absorption bands in the nitrate and sulphate spectra share practically the same position, while in the case of the chloride, there is a clear bathochromic shift. The difference between these spectra also occurs in the range of greatest transmittance, which in the case of the chloride spectrum is located at approximately 520 nm, and in the case of other Cr compounds, at approximately 480 nm.

The differences outlined above can be explained, at least in part, by reference to differences in the structure of the water solutions of the compounds under discussion and the form of the constituent complex ions. In the case of chromium trichloride, it is known [50] that $[Cr(H_2O)_5Cl]^{2+}$ and $[Cr(H_2O)_4Cl_2]^+$ may occur alongside $[Cr(H_2O)_6]^{3+}$ in aqueous solutions. In simulation studies we found replacement of one of the waters of $[Cr(H_2O)_6]^{3+}$ by chloride to cause a purplish pink to green colour change, but that introduction of a second chloride causes negligible further change in colour; di- and mono-chloro-complexes are the same colour. The essentially constant green colour of $CrCl_3$ solutions in the concentration range 0.02 to 0.30 mol dm^{-3} probably conceals a varying proportion of the two chloro-complexes. The straight-line chromaticity trace determines the dominant wavelength as 525 nm.

In solutions of the two other salts, the concentration of hexaaquaions is probably greater or even prevalent. However, the cause of the occurrence of the classic chromaticity trace indicating dichromatism, where the maximum excitation purity occurs for a definite concentration, remains unclear. As has been mentioned,

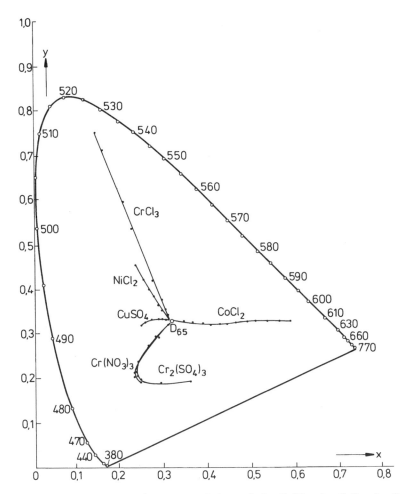

Figure 3.25 Chromaticity traces of aqueous solutions of: $1 - CuSO_4$, $2 - CrCl_3$, $3 - Cr_2(SO_4)_3$, $4 - Cr(NO_3)_3$, $5 - CoCl_2$, and $6 - NiCl_2$ (dependence on concentration).

Griffiths [56] believes that the nature of the curve is the result of purely physical factors (in a discussion of such a situation in the case of organic dyes). Some aspects of this question were discussed in detail by Zausznica [54].

A more thorough analysis of the data in Table 3.21 reveals that in the case of $CrCl_3$, an increase of the concentration of the compound causes a systematic decrease of the value of the x coordinate and an increase of the value of the y coordinate, which results in a negative slope of the straight line in the chromaticity diagram.

Let us now consider the chromaticity coordinates for $NiCl_2$. As can be seen in Figure 3.25, all aqueous solutions (except for the most dilute) are green, and the dependence of the coordinates on concentration is also a straight line.

Comparing $CrCl_3$ and $NiCl_2$ one may conclude that in both cases the green colour is chiefly related to the absorption minimum at approximately 500 nm (which,

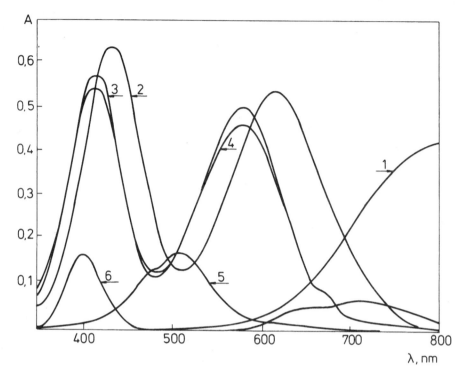

Figure 3.26   Absorption spectra of aqueous solutions of the compounds shown in Figure 3.25.

however, is usually much wider for Ni(II) compounds) and to the position of two absorption bands. For the former compound, the maxima are at 390 and 600 nm; for the latter at 400 and 700 nm. The difference between these metal ions is mostly due to the different charge of the metal ion, as a result of which the g parameter for Cr^{3+} is approximately $17\,400\,cm^{-1}$, while for Ni^{2+} it is $8000\,cm^{-1}$. In the spectrum of the Cr^{3+}, the first long-wave band in fact corresponds to the second band in the Ni^{2+} spectrum (for the same ligands). Finally, the difference in terms of coordination chemistry is due to the fact that in a solution of $NiCl_2$ the prevailing form is the hexaaquaion rather than mixed complexes.

3.3.2   Colour of transition metal compounds in the solid phase and in solutions

So far in our discussion of the colour of transition metal compounds, no attention was paid to the special role of the phase whose colour is determined for instance in studies of absorption or reflectance spectra. The kind of interaction between the molecules of a chemical compound and the environment (matrix) undoubtedly depends on the kind of phase, which results in a change of energy levels and of the probability of electronic transitions between them. Thus a change of the environment from the gaseous phase, where such interactions can be neglected, to the liquid phase (solution) or the solid phase, where structural aspects play a significant role, may bring about a more or less pronounced change in colour.

We shall consider three main aspects here: medium effects (especially of organic solvents), the effects of temperature and, briefly, of pressure, and some aspects of colour in the solid phase.

3.3.2.1   *Solvatochromism*

The term **solvatochromism** [57] is used, understood, and interpreted in different and often inconsistent ways. The very name indicates that the solvent causes the occurrence or change of colour of the solute. In the most fundamental terms the change should be in relation to the gas phase, and indeed some thermodynamic quantities, for example enthalpies of transfer from the gaseous phase into the liquid phase, have been calculated. $\{Mo(CO)_4\}_2(bipym)$ provides one of the very rare instances where solvatochromic shifts from the solid state into solution are known, for its diffuse reflectance spectrum in an MgO matrix has been recorded [58]. Extension to surfaces and the solid phase may be illustrated by the use of solvatochromism to probe aerosil and suprasil surfaces – untreated, chlorinated, and epoxide-coated [59] – and the cages and channels in zeolite structures [60]. However, in most cases solvatochromism experiments and discussions deal with colour changes on transfer between solvents (including supercritical fluids [61]).

Several experimental methods have been used in studies of solvatochromism, taking this term in its widest sense. These include infrared and Raman spectroscopy, NMR and ESR spectroscopy, and thermochemistry, as well as ultraviolet-visible spectroscopy. Compounds of the $Mo(CO)_4(diimine)$ type provide good illustrations of the study of solvatochromism in the widest sense, including vibrational and NMR spectra, and thermochemistry, as well as electronic spectra [62]. However if solvatochromism is understood as specifically involving a change in colour, as in the present book it should, then of course the direct spectroscopic method of examining electronic spectra in the visible region is the relevant technique. As has been emphasised earlier, the spectrum curve provides only very approximate grounds for the prediction of colour, and colour changes, on the basis of absorption maxima and minima, and of their shifts. Nonetheless the absorption spectrum provides the basis for the determination of the chromaticity coordinates, x, y, and Y, which enable full characterisation of colour in the CIE system (cf. Chapter 1.5), as well as for the calculation of other values used in practice, such as the hue angle (cf. Chapter 5). Until very recently this approach has not been used in solvatochromic studies. The work of one of the present authors and his coworkers [55, 63, 64] is among the first in this field and will be discussed below in some detail.

In publications devoted to the evaluation of solvatochromism from electronic spectra, attempts are usually made to express transition energies (expressed in wavenumbers) corresponding to the absorption band in question in terms of a correlation with one or more parameters characterising the solvents. However there are generally several absorption bands in the electronic spectra of transition metal complexes, and the actual transition energies can only be accurately determined by a resolution of the spectrum into its component bands. A further complication is that change of solvent often causes not only a change of transition energies but also changes in intensities and half-widths of individual bands. All these quantities have a definite physical basis and a full evaluation of solvent-solute interactions should analyse all such changes. Quantitative assessment of colour changes may be valuable in the interpretation of solvatochromism. If the observed change in colour is slight, but visibly apparent, then the chromaticity coordinates may provide additional important information concerning the changes occurring in the system under investigation.

When considering solvatochromism of transition metal complexes and compounds, three different aspects must be considered:

(a) solvatochromism of *d-d* transitions and of charge-transfer transitions;

(b) specific and non-specific interactions;

(c) interactions connected with a change in the inner coordination sphere and solvatochromism resulting from interaction with the outer coordination sphere.

These issues are not completely independent of each other, but in the literature of solvatochromism interpretation is often restricted to only one of these inter-related aspects.

In the case of solvatochromism of ligand field and charge transfer transitions, it is usually assumed [8] that ligand field transition energies do not change when the solvent is changed. However, more detailed studies reveal that these energies do change, even though the changes are often very small. It is not easy to estimate this effect in the absorption spectra of solutions as bands overlap, requiring the use of computer deconvolution of the spectrum envelope to obtain precise evaluation of the solvent effect.

One of the models often used to evaluate this solvent effect is Onsager's reaction field model [65, 66]. Bartecki and Stelmaszek [67] put forward the new hypothesis that in the reaction field model, which assumes the formation of a cavity containing molecules of the solute, the symmetry of the molecule of the complex should be taken into account. Working with the crystal field model, they obtained the following expression for octahedral complexes:

$$Dq^{\text{eff}} = Dq^0 + \text{const}\frac{5(\varepsilon - 1)}{5\varepsilon + 4}; \quad \text{const} = Dq^0(a/a_0)^9,$$

where a is the distance of the central ion from the ligand, a_0 is the radius of the cavity, Dq^0 denotes the crystal field parameter for the gaseous phase (without interaction), Dq^{eff} is the effective value, including the additional stabilisation caused by the nonspecific interaction with the solvent with a macroscopic permittivity ε.

The dependence was tested for fourteen $[Ni(H_2O)_6]Cl_2$ systems in organic and mixed solvents [67]. The results are given in Table 3.30, and the dependence $Dq^{\text{eff}} = f[5(\varepsilon - 1)/(5\varepsilon + 4)] \cdot 10^3$, in Figure 3.27.

The correlation coefficient of 0.97 indicates that the above dependence is a good representation of the effect of nonspecific interactions on the value of the effective field strength Dq^{eff}. It should be stressed that the solvents used: water, ethanol, methanol, and their mixtures have a very similar effect. Consequently, in practice the $10Dq$ values calculated on the basis of the values of two spin-allowed transitions

Table 3.30 Dq^{eff} and B values of $[Ni(H_2O)_6]^{2+}$ ion as dependent on dielectric permittivity of organic and mixed solvents [67]

Sovent	ε	$\dfrac{5(\varepsilon-1)}{5\varepsilon+4}$	Dq^{eff} (cm^{-1})	B (cm^{-1})
Water (W)	79.6	0.9776	825.2	972.2
79%W + 21%M	69.9	0.9745	821.2	961.8
68%W + 32%E	60.7	0.9707	821.4	964.5
43%W + 57%M	52.5	0.9662	818.3	961.8
43%W + 57%E	46.0	0.9615	815.5	965.0
15%W + 85%M	38.8	0.9545	807.2	963.2
Methanol (M)	33.2	0.9471	789.5	928.6
86%M + 14%E	32.2	0.9455	795.7	951.3
75%M + 25%E	31.5	0.9443	795.3	946.0
15%M + 85%E	31.2	0.9438	801.0	970.6
50%M + 50%E	29.8	0.9411	789.5	942.4
25%M + 75%E	28.1	0.9376	788.5	933.3
14%M + 86%E	27.3	0.9360	780.4	938.2
Ethanol (E)	24.8	0.9296	774.4	928.0

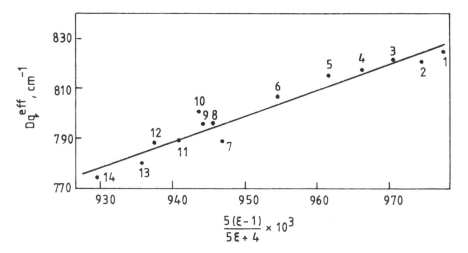

Figure 3.27 Dependence of Dq^{eff} values for Ni^{2+} solvento-complexes on the expression $(5\{\varepsilon-1\})/(5\varepsilon+4).10^3$ for 14 organic and mixed solvents; ε denotes dielectric permittivity. Solvent numbers from 1 to 14 are in descending order in Table 3.30 [67].

change only slightly, from 7740 to 8250 cm^{-1}, i.e. by some 500 cm^{-1}. It can be easily verified that the corresponding colours differ very little from each other and are all practically green.

The dependence was also tested for chromium(III) nitrate hexahydrate in the following organic solvents: water, methanol, ethanol, t-butanol, and some binary mixtures of these solvents [68]. The results show a high correlation between Dq^{eff} and the expression $5(\varepsilon-1)/(\varepsilon+4)$; $r=0.986$. The following values were obtained:

Table 3.31 Wavenumbers and oscillator strengths of two spin-allowed bands in the absorption spectra of $Cr(NO_3)_3 \cdot 6H_2O$ in non-aqueous solvents [68]

Solvent	ν_1 (cm^{-1})	f''^a	ν_2 (cm^{-1})	f''^a
Water (W)	17392	1.79	24426	3.19
W + M (1 : 1)	17353	1.72	24450	3.20
W + E (1 : 1)	17342	1.77	24409	3.32
W + t − B (1 : 1)	17336	1.72	24419	3.10
Methanol (M)	17306	1.59	24379	2.89
E + M (1 : 1)	17268	1.93	24287	3.31
E + M (1 : 1)	17259	2.09	24133	3.54
Ethanol (E)	17243	2.22	24132	3.88
t − B + M (1 : 1)	17223	2.06	24233	3.51
t − B + E (1 : 1)	17165	2.40	24089	4.22
t − B + E (4 : 1)	17122	2.59	23941	4.67
t − Butanol (t−B)	16984	2.69	23803	4.86

^{a} Oscillator strength, in 10^{-4} units.

$Dq^0 = 1416 \pm 19\,\mathrm{cm}^{-1}$ and $a/a_0 = 0.85 \pm 0.02$. Table 3.31 presents the wavenumbers and oscillator strengths for two spin-allowed bands obtained as a result of a computer resolution of spectra.

As can be seen, the difference between the energies of the two spin-allowed transitions obtained in the solvents studied is several hundred cm^{-1} (while the difference $\nu_2 - \nu_1$ is almost constant at about 7000 cm^{-1}). Now it is clear why the effect of solvatochromism is usually neglected, especially for one class of solvents, such as alcohols in this case.

The oscillator strengths for both transitions were calculated from the dependence proposed in papers [69, 70], also based on Onsager's reaction field model:

$$f''/f^0 = [s(n^2 - 1) + 1]^2/n,$$

or in a rearranged form:

$$(nf'')^{1/2} = (f^0)^{1/2}s(n^2 - 1) + (f^0)^{1/2},$$

In the above expressions, n is the refractive index[6] of the solvent, s is the shape parameter, f^0 is the oscillator strength of a hypothetical solvent whose refractive index is equal to 1.

As can be seen, the experimental oscillator strengths f'' (in the order of 10^{-4}) in different solvents are markedly different and it seems that the intensities of both bands in other solvents manifesting only nonspecific interaction can be estimated on this basis. However, the obtained values of the s parameter are greater than 1, which is the maximum value theoretically predicted in the model. This is probably due to an additional effect of the ligand field, which was not taken into account in paper [70].

---

[6] For the sodium D line.

In subsequent studies of the solvent effect, Staszak and Bartecki [71] showed that studies of solvatochromism in chemically different solvents should also take into account specific effects, which can be represented by including the donor number (DN) or acceptor number (AN), introduced by Gutmann *et al.* [72]. In [71] the following solvents were used: water, DMSO, DMF, methanol, ethanol, acetone, and THF, that is solvents with different physical and chemical properties. It was found that the position of the absorption bands obtained from the computer resolution and the Dq^{eff} values calculated on that basis satisfy the equation

$$Dq^{eff} = Dq^0 + \text{const}[5(\varepsilon - 1)/(5\varepsilon + 4)] + cDN$$

both for $Cr(NO_3)_3 \cdot 6H_2O$ and for $K_3Cr(NCS)_6 \cdot 4H_2O$.

The question of donor-acceptor interactions, their effects on the phenomenon of solvatochromism, and their relation to various polarity scales, have been dealt with at length [73]. Two main types of solvatochromism should be recognised in transition metal complexes [74]. These are:

(a)   solvatochromism caused by structural changes in the inner coordination sphere, dependent on the DN, and

(b)   solvatochromism caused by solute-solvent interaction in the outer coordination sphere, generally dependent on the AN of the solvent.

The classic example of the former is $[Cu(acac)(tmen)]^+$ [75], where acac = acetylacetonate (2,4-pentanedionate) and tmen = tetramethyl-1,2-ethanediamine. Here the solvent has access to the copper in the vacant fifth and sixth coordination positions, and solvent donor properties are believed to be the sole determinant [76]. The classic and most studied example of the latter is $[Fe(bipy)_2(CN)_2]$, discovered by Barbieri [77] and later fully investigated by Schilt [78]. Here solvation is dominated by hydrogen-bonding interactions, and it is the AN of the solvent which determines solvatochromic shifts. In view of the relative scarcity of AN values, it is also customary to seek correlations with transition energies for Reichardt's $E_T(30)$ dye (Figure 3.28). This is a versatile (except for its inconveniently low solubility in water)

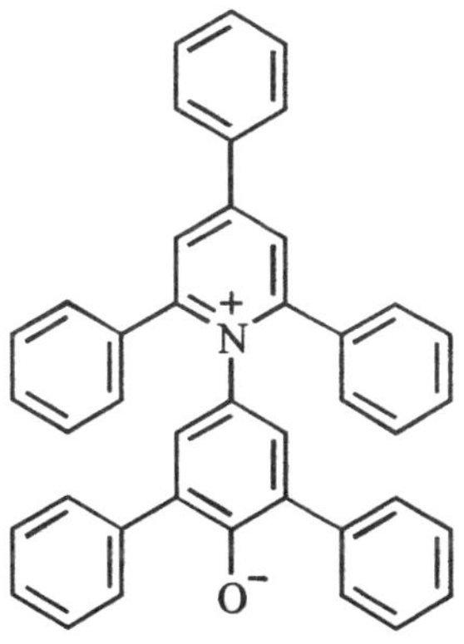

Figure 3.28   Reichardt's E_T {strictly $E_T(30)$} polarity indicator.

and very sensitive probe of solvation when dominated by acceptor properties and hydrogen-bonding [79, 80]. Both donor and acceptor properties are important in determining the solvatochromic behaviour of a few complexes – an example of such a manganese complex appears in Section 3.3.2.2 below.

In solvatochromic studies the observed colours and their changes with solvent nature are seldom described in detail. However such changes have been described for some iron(II)-diimine-cyanide complexes [81, 82]. These diamagnetic low-spin (t_{2g}^6) ternary complexes give strongly coloured solutions in water, with absorption spectra rather similar to their tris-diimine parent complexes. But, as Schilt first reported, in contrast to the tris-diimine complexes, the ternary species manifest strong solvatochromism, with the colours of their solutions strongly correlated with the polarity of the solvent. These issues will be discussed in more detail below, and also in Chapter 5.

3.3.2.2 Solvatochromic inorganic species

A number of solvatochromic inorganic species have been mentioned briefly in the course of the above discussion. In this section we shall summarise the groups of inorganic complexes and organometallic compounds which show marked solvato-chromic properties. Many are ternary species of the metal-diimine-cyanide or metal-diimine-carbonyl type, with solvatochromic behaviour residing in one or more intense (ε often in the region of 10^4 to 10^5) MLCT (metal-to-ligand charge-transfer) bands. Metals with the low-spin d^6, i.e. t_{2g}^6, configuration are particularly able to add some synergic π back-bonding, in other words MLCT, to complement and augment their σ-bonding. Solvatochromic LMCT bands are also known, though LMCT is less commonly encountered and gives less intense bands. There are occasional reports of ligand-to-ligand charge-transfer (LLCT) through appropriate metal orbitals, while the solvatochromism of metal-to-metal charge-transfer bands of mixed valence compounds can give valuable kinetic information on electron transfer (Section 3.3.3). Many solvatochromic inorganic species have been described, of varying nature and sensitivity to solvent variation [83], but as yet no species as sensitive as Reichardt's solvatochromic organic compounds has been reported.

The first group to mention comprises iron(II)-diimine-cyanide complexes, of which the paradigm is $Fe(bipy)_2(CN)_2$ (cf. above), though in fact the first report of marked solvatochromic behaviour for an inorganic complex was that described for $Fe(phen)_2(CN)_2$ [84]. Correlations of wavenumbers of maximum absorption, or transition energies, with solvent AN values have been reported for $Fe(bipy)_2(CN)_2$ [81] and for $Fe(phen)_2(CN)_2$ [85]. Such plots actually consist of two slightly separated lines, for hydroxylic and for non-hydroxylic solvents, with points for chlorohydrocarbon solvents of intermediate properties between the two lines – as shown in the plot of ν_{max} against solvent AN values for the extensively documented complex $Fe(phen)_2(CN)_2$ [85, 86] in Figure 3.29.

We shall deal with the use of $Fe(phen)_2(CN)_2$ as an indicator for acceptor numbers, and document its colour and chromaticity, in Section 5.4. Direct comparison of this type of complex with standard organic dyes may be made by noting the ranges of 555 nm (18 020 cm^{-1}) to 629 nm (15 900 cm^{-1}) for $Fe(bipy)_2(CN)_2$ and of 515 nm

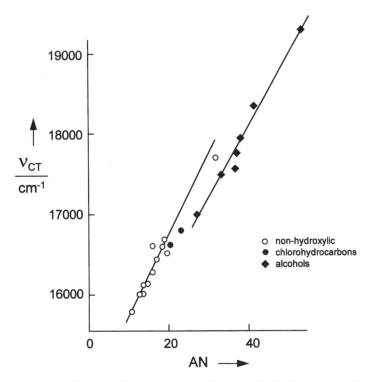

Figure 3.29 Correlation of wavenumbers of maximum absorption for the lowest energy charge transfer band (ν_{CT}) of Fe(phen)$_2$(CN)$_2$ with solvent acceptor numbers, AN.

(19 420 cm^{-1}) to 710 nm (14 080 cm^{-1}) for Reichardt's E_T(30) betaine, for the solvent range from methanol to pyridine. The considerably lower solvent sensitivity of this type of iron-diimine-cyanide complex in comparison with Reichardt's E_T(30) betaine is also apparent from plots correlating these parameters [87].

One of the great advantages of the iron-diimine-cyanide group of complexes is the possibility of tailoring colour, solvent sensitivity, and solubility behaviour by diimine ligand variation. Ligands may be of the phen/bipy type, including such variants as bipym and btz, they may be diazabutadienes, or they may be Schiff bases derived from pyridine 2-carboxaldehyde, 2-acetyl pyridine, or 2-benzoyl pyridine (see Figure 3.30 for formulae and abbreviations). Solubility in hydrocarbon solvents can be favoured by phenyl or bulky alkyl substituents; water-solubility can be conferred or enhanced by sulphonate or carboxylate substituents – or by switching from uncharged Fe(diimine)$_2$(CN)$_2$ to alkali metal salts of [Fe(diimine)$_2$(CN)$_4$]$^{2-}$ anions, or indeed of iron(III) analogues [Fe(diimine)$_2$(CN)$_2$]$^+$ or [Fe(diimine)(CN)$_4$]$^-$. Some of these ternary Schiff base-cyanide complexes are perhaps the most obviously solvatochromic to the eye, in that the colours of several of them span the visible range from red through to blue, as depicted in Figure C3. Although qualitatively the perceived colour range for these Fe(diimine)$_2$(CN)$_2$ complexes suggests that they are roughly as sensitive to their solvation environment as Reichardt's betaine dyes,

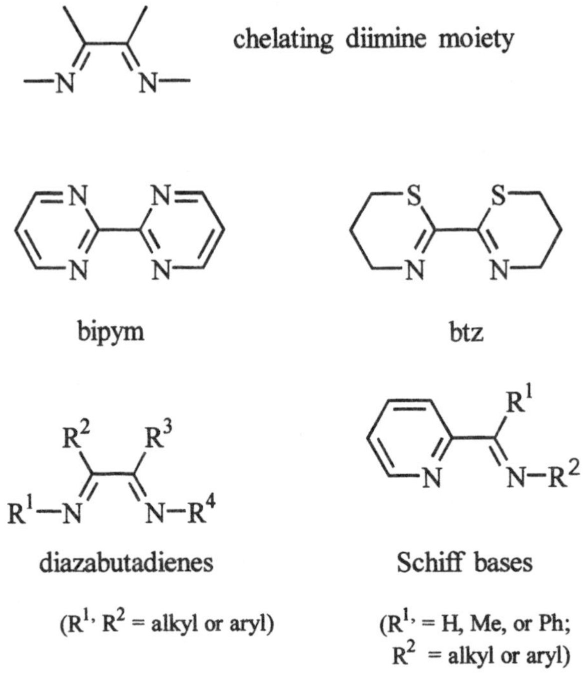

chelating diimine moiety

bipym

btz

diazabutadienes

$(R^1, R^2 = \text{alkyl or aryl})$

Schiff bases

$(R^1, = \text{H, Me, or Ph};$
$R^2 = \text{alkyl or aryl})$

Figure 3.30   Formulae for some ligands containing the chelating diimine group.

quantitative comparisons indicate that in fact they are considerably less sensitive. Indeed both the Fe(Schiff base)$_2$(CN)$_2$ and the Fe(diazabutadiene)$_2$(CN)$_2$ series of complexes are significantly less solvatochromic than Fe(bipy)$_2$(CN)$_2$ [88].

Iron-diimine-cyanide complexes have proved valuable in solvatochromic investigations of solvation in mixed solvents [89, 90], salt solutions [91], and micelles [92]. Their use has permitted the estimation of acceptor numbers for some of these media.

The next important group are the Group 6-tetracarbonyl-diimines, especially of molybdenum, to the study of which much work has been devoted [93]. Colour variations between solutions of these compounds in a range of solvents, which generally range over various shades of red, orange, and yellow, are clearly apparent to the eye, though they appear less marked than for ternary iron(II)-diimine-cyanide complexes. Figure 3.31 shows the relation between ν_{max} and solvent acceptor properties, in the form of E$_T$(30) values, for Mo(CO)$_4$(bipy). The two lines again correspond to hydroxylic and non-hydroxylic solvents, with one or two indeterminate intermediate solvents such as chloroform in between. The separation between the two lines, small but significant for the iron complex in Figure 3.29 above, is dramatic for the molybdenum compound. As for the iron-diimine-cyanide complexes, the Group 6-diimine-carbonyl probes can be tailored by appropriate ligand variation, with the additional possibility of Cr, Mo, W variation {very little is known of the solvatochromic properties of Ru(bipy)$_2$(CN)$_2$ [94], nothing of those of Os(bipy)$_2$(CN)$_2$}.

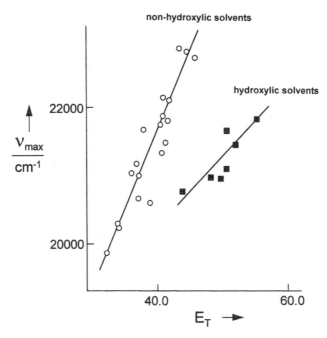

Figure 3.31 Correlation of wavenumbers of maximum absorption (ν_{max}) for the lowest energy charge transfer band of $Mo(CO)_4(bipy)$ with solvent E_T values.

Molybdenum-diimine-carbonyl compounds have been used to assess solvation in non-aqueous solvent mixtures [89] and in microemulsions [95]. The often complementary solubility characteristics of $Fe(diimine)_2(CN)_2$ and of $Mo(CO)_4(diimine)$ derivatives can sometimes permit monitoring a wider range of solvents than either alone, as in oil/water microemulsion systems.

The Group 7 analogues of $M(CO)_4(diimine)$ are carbonyl-halide-diimine compounds $M(CO)_3(diimine)X$, well-established for M = Mn, Tc, and Re. $Mn(CO)_3$-(bipy)Br is some 50% more solvent sensitive than $Fe(bipy)_2(CN)_2$ in hydroxylic solvents, but of comparable solvent sensitivity in non-hydroxylic media [96]. Clearly the bromide ligand is having an effect on solvation here. Strongly solvatochromic MLCT transitions have also been observed in the analogous organometallic rhenium(I) compounds $Re(phen)(CO)_3Cl$ and phen-substituted derivatives [97], *fac*-[$Re(bipy)(CO)_3H$], and *fac*-[$Re(bipy)(CO)_3Ph$]. Substantial shifts of the maxima for the last two compounds have been noted, up to 50 nm (from 22 000 to 24 000 cm^{-1}), which is manifested in a change of colour from yellow to orange. The solvatochromic behaviour here has been interpreted using dielectric continuum theory.

Ternary molybdenum- and tungsten-diimine-cyanide complexes link the two major groups of complexes discussed so far. Despite the high oxidation state, 4+, here the solvatochromic bands are still MLCT in character. Wavenumbers of maximum absorption for the [$W(bipy)(CN)_6$]$^{2-}$ anion correlate well with respective values for $Fe(bipy)_2(CN)_2$, both in hydroxylic and non-hydroxylic solvents. The solvent sensitivity of the tungsten complex is 1.2 relative to $Fe(bipy)_2(CN)_2$ [98].

Moving slightly away from diimines, two related groups are the pentacyanoferrates $[Fe(CN)_5L]^{n-}$ and their Group 6 carbonyl analogues $M(CO)_5L$. The colours of pentacyanoferrate(II) complexes range from the intense deep blue of the N-methylpyrazinium complex through deep red for the pyrazine complex to pale yellow for several of the (substituted) pyridine derivatives. These colours reflect the large effect of L on the wavelength of maximum absorption for the main MLCT band (λ_{MLCT}), which in aqueous solution is at 659 nm for the N-methyl-pyrazinium complex, 458 for L = pyrazine, and down to 345 nm for L = 4-methyl-pyridine (γ-picoline). The complexes are solvatochromic [83, 99], to degrees which depend greatly on the nature of the group L. Thus the change in λ_{MLCT} on going from aqueous solution into methanol is +68 nm for L = N-methyl-pyrazinium, +49 nm for L = pyrazine, but just −2 nm for phenazine. In terms of transition energies, the biggest shift is for L = 4-t-butyl ($-2540\,cm^{-1}$), with a shift of $-1410\,cm^{-1}$ for L = N-methylpyrazinium being almost exactly equal to that for $Fe(bipy)_2(CN)_2$. Solvatochromic shifts for pentacyanoferrates(III) are generally in the opposite direction from those for the pentacyanoferrates(II), as one would expect in view of the change from MLCT to LMCT. The shifts for the $[Fe^{III}(CN)_5L]^{n-}$ complexes correlate with such solvent parameters as Reichardt's E_T, Kosower's Z, and Gutmann's acceptor numbers (AN). It is said that in all these cases the ability of the solute's dipole to polarise the solvent is significant [100]. The solvatochromic properties of various pentacyanoferrates have proved useful in probing solvation in salt solutions [101], in mixed solvents [100, 102, 103], and in micelles [104].

The solvatochromic behaviour of several five-coordinate manganese(III) complexes Mn(LLLL)(NCS), where LLLL is one of the tetraazamacrocycles shown in Figure 3.32 [105], is of particular interest as it reflects both donor and acceptor properties [106]. Although the coordination number of the manganese is formally five, it appears that in solution there can be significant interaction with a molecule of donor solvent at the sixth coordination site. Complementarily, there is the possibility of hydrogen-bonding at the thiocyanate ligand.

One of the rare five-coordinate complexes of cobalt(III), viz. the cation $[Co(Me_4cyclam)(NCS)]^+$ {see Figure 3.33 for $Me_4cyclam$ formula} links, via the closely

	R^1	R^2	R^3
LLLL^1	H	H	H
LLLL^2	H	Me	H
LLLL^3	H	H	4-tolyl

Figure 3.32 Formulae for the substituted tetra-azamacrocycles LLLL^n in solvatochromic complexes Mn(LLLL^n) (NCS).

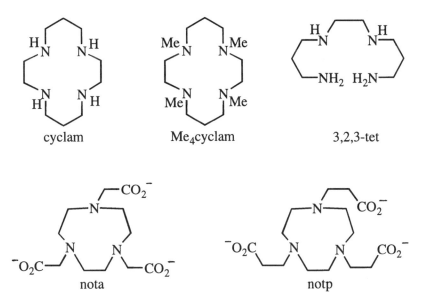

Figure 3.33 Formulae and abbreviations for azamacrocyclic ligands which may give substitution-inert complexes.

related but octahedral complex [Co(cyclam)(NCS)$_2$]$^+$ [107], to other Werner complexes. Several of these, both of cobalt(III) and of chromium(III), were found to show slight solvatochromism of their d-d bands in the first report on inorganic solvatochromism [84]. Complexes which showed significant changes in λ_{max} or ε on transfer from water into methanol, and in some cases also into nitromethane and acetone, included *cis*-[Co(en)$_2$Cl$_2$]$^+$, Co(NH$_3$)$_3$(NO$_2$)$_3$, and even the highly symmetrical [Cr(NCS)$_6$]$^{3-}$. Later workers reported further data on the [Cr(NCS)$_6$]$^{3-}$ anion [108], and described solvatochromism for [Co(CN)$_6$]$^{3-}$ [109], for [Cr(edta)]$^-$ and [Co(edta)]$^-$, where a tolerably good correlation with acceptor numbers was reported, and for *trans*-[Cr(en)$_2$F$_2$]$^+$. Hydrogen-bonding interactions with the fluoride ligands must clearly be important for the last-named. They also dominate the solvatochromic behaviour of macrocyclic analogues *trans*-[Cr(3,2,3-tet)X$_2$]$^+$ {the ligand 3,2,3-tet is shown in Figure 3.33; X = e.g. F, Cl, Br, CN, NCS}; solvation at the ligands X affects the metal–ligand σ-bonding. There is a correlation of ν_{max} with solvent acceptor numbers [110]. Investigations of cobalt(III) complexes have also included some with macrocyclic ligands. The colours of solutions of the triazatricarboxylate complexes Co(nota)(H$_2$O)$_3$ and Co(notp)(H$_2$O)$_3$ {formulae for nota and for its propionate analogue notp are shown in Figure 3.33} [74] showed a gradation from blue purple to reddish purple, corresponding to wavenumber changes over the ranges 19 000 to 20 000 cm^{-1} and 26 000 to 27 000 cm^{-1} for the two absorption maxima. Values of ν_{max} for the former band again correlated with solvent acceptor numbers (AN), as did those for [Co(CN)$_6$]$^{3-}$.

There have been few reports of solvatochromism for square-planar complexes. The same is true for organometallic compounds other than the Group 6-diimine-carbonyls, though we should mention the somewhat similar compound [Ti(cp)$_2$(NCS)$_2$] [98] here.

3.3.2.3   Thermochromism

Thermochromism denotes a change in the colour of a substance as a result of a change of temperature.[7] Such changes may occur in all phases. Detailed studies of thermochromism may provide information about mechanisms and products of reactions, structural changes, the forms of complex ions, and several other aspects of transition metal chemistry. This section deals mainly with thermochromism of solid transition metal compounds, but includes a brief mention of solution chemistry towards the end.

Firstly we review briefly the dependence of electronic spectra of transition metal compounds on temperature. Without going into details, it can be said that changes of temperature result in changes of the intensity and energy of electronic transitions, which are different for spin-allowed and spin-forbidden transitions. This is discussed in a concise and straightforward way elsewhere [111]. As regards changes in the intensity of electronic transitions, it has been concluded that oscillator strength does not depend on temperature, so it can be assumed that its changes have no effect on the form and parameters of the electronic band. However, as a result of an increase of temperature, in addition to the 0-0 transition (or a progression based on this transition), there are also transitions involving the population of the higher oscillation levels of the ground state, e.g. 1,1 or 2,2. In such cases the intensity distribution shifts towards longer wavelengths, which at the same time results in decrease of the intensity of the main band. If a very good separation is obtained, the bands are observed, otherwise the absorption band is deformed. Thus some change of colour may be anticipated as a result of changes of absorption and transmission caused by temperature changes.

In the case of electronic-forbidden transitions, such as *d-d* transitions, their occurrence is explained by electronic-oscillation interaction, i.e. those transitions are allowed as vibronic transitions. Their oscillator strengths usually decrease with the decrease of temperature, but the problem is not straightforward [111].

As is well-known, a more important parameter determining changes of colour is the change of the position of absorption bands and the accompanying shift of the spectrum minima. It follows from the equation defining the Dq parameter that this value depends on a^{-5}, where a is the distance from the ligand to the central ion. The change may be the result of oscillation, so a decrease of temperature causes a decrease of the population of higher oscillation states and an increase of Dq. However the change of the position of electronic transition maxima depends on the slope of the excited state curve in the graph of $E = f(Dq)$. If $dE/dQ > 0$, a hypsochromic shift occurs (towards shorter wavelengths); if $dE/dQ < 0$, a bathochromic shift takes place. It follows that the effect of temperature changes on colour cannot be predicted if such information is not available.

All the above aspects should be taken into account when considering the thermochromism of transition metal compounds, but physicochemical and chemical changes, which cause noticeable changes of colour, are of fundamental significance.

---

[7] The thermochromism of $Ag_2[HgI_4]$ and of $Cu_2[HgI_4]$ is illustrated in Colour Plate 29 of H.W. Roesky and K. Möckel, *Chemical Curiosities*, VCH Weinheim, 1996.

Table 3.32 Thermochromic data for some cobalt(II) and cobalt(III) compounds

		60 °C		170 °C		235 °C	
$CoSO_4 \cdot 7H_2O$	dark red	$\longrightarrow$ I	lilac	$\longrightarrow$ I	blue	$\longrightarrow$ I	dark red purple
		162 °C		195 °C			
$CoCO_3 \cdot 6H_2O$	purple	$\longrightarrow$ I, III	grey	$\longrightarrow$ III	black		
		120 °C		170 °C		230 °C	
$[Co(NH_3)_5Cl]Cl_2$	red pink	$\longrightarrow$ III	purple	$\longrightarrow$ III	turquoise	$\longrightarrow$ III	black

Generally we can distinguish reversible and irreversible thermochromism. [25] is a very useful monograph on this subject.

In the case of irreversible thermochromism, a change of colour and of the spectrum point to a specific change of, for instance, the crystallographic structure, the coordination number, and the symmetry of the complex, a change of configuration, e.g. from high-spin to low-spin, etc. At the same time it means that a return to the initial temperature does not bring back the original colour. Many studies have shown that slow heating of transition metal complexes may cause noticeable colour changes. Some examples from the chemistry of cobalt(II) and of cobalt(III) are given in Table 3.32 [112]. On the basis of analysis of results for eight compounds, colour changes can be attributed to three causes (the numbers I, II, and III correspond to their use in Table 3.32):

(I)   The loss of water of crystallisation, usually in the range 50 to 150 °C. For cobalt compounds this is often accompanied by a red/purple to blue colour change).

(II) Oxidation. For cobalt(II) compounds this gives a change of colour to black Co_3O_4 at about 230 °C or above.

(III) Various thermal decomposition processes, such as loss of ammonia or of carbon monoxide.

Irreversible dehydration of various salts or transition metal complexes is a commonly enough encountered process in chemistry. One classic example is the dehydration of $CuSO_4 \cdot 5H_2O$ (a process described in standard chemistry textbooks). As is well known, when blue vitriol is heated, the blue crystals change into the white powder of anhydrous $CuSO_4$. If even a trace of water is present (e.g. contained in organic solvents), the blue colour returns.

As the blue colour of $CuSO_4 \cdot 5H_2O$ is connected with the d-d band in the Cu(II) ion (broad and asymmetrical as in many other Cu(II) compounds), it is interesting to consider this transition in the anhydrous compound. It turns out that gradual dehydration, through the stages of trihydrated and monohydrated sulphate, to anhydrous sulphate causes a shift of the band from approximately 750 to 800 nm. The d-d band does not disappear (let us recall here that the aqueous solution of the original salt reveals a band at this wavelength). The authors of [25] claim that the colourlessness of the anhydrous salt is an apparent phenomenon due to pulverisation and the shifting of the spectrum towards infrared. However, this explanation does not seem to be sufficient. A bathochromic shift of the spectrum also causes an increase of the transmission range, i.e. practically all light is transmitted through the sample and does not cause the formation of colour. Of course, the d-d transition still occurs.

Interesting information on colour changes during dehydration of cobalt(II) and nickel(II) complexes $[M(H_2O)_6]X_2 \cdot nH_2O \cdot 2L$, where $L = $ hmta (hexamethylenetetramine), has been published [113]. For the cobalt complexes:

1. $[Co(H_2O)_6]Cl_2 \cdot 4H_2O \cdot 2L$ $\xrightarrow{50-130\,°C}$ $[CoCl_2L_2] + 10H_2O$

     pink   blue

2. $[Co(H_2O)_6]Br_2 \cdot 3H_2O \cdot 2L$ $\xrightarrow{55-105\,°C}$ $[CoBr_2L_2] + 9H_2O$

     pink   blue

3. $[Co(H_2O)_6]I_2 \cdot 2H_2O \cdot 2L$ $\xrightarrow{55-110\,°C}$ $[CoI_2L_2] + 8H_2O$

     pink   green

4. $[Co(NCS)_2(H_2O)_4] \cdot 2L$ $\xrightarrow{85-125\,°C}$ $[Co(NCS)_2L_2] + 4H_2O$

     orange   blue purple

In the above examples, and similarly in Ni(II) complexes, there is a clear change of colour caused by a change of the geometrical structure of the complex from octahedral or tetragonal to tetrahedral.

The thermal properties of transition metal ammine complexes were thoroughly studied many years ago by Wendlandt and Smith [114]. Their book also contains some data on the colour of the complexes.

Table 3.33 presents the changes of colour in some Co(III) complexes, and Figure 3.34 shows the reflectance spectra of $[Co(NH_3)_5(H_2O)]Cl_3$.

As can be seen, three groups of spectra can be distinguished: the first at 25 and 50 °C, the second at 75 and 125 °C, and the third at 150, 175, and 222 °C. In the first case the complex is not changed, in the second case the authors assume the formation

Table 3.33 Colour changes observed during heating some Co(III) complexes [114]

Complex	Colour change	Temp. range (°C)
$[Co(NH_3)_6]Cl_3$	orange → blue	235–237
$[Co(NH_3)_5Cl]Cl_2$	purplish pink → blue	213–215
$[Co(NH_3)_4Cl_2]Cl$	green → blue	180–182
$[Co(NH_3)_3Cl_3]$	dark green → blue	155–160
$[Co(NH_3)_4(H_2O)_2]Cl_3$	purplish grey → blue	167–170

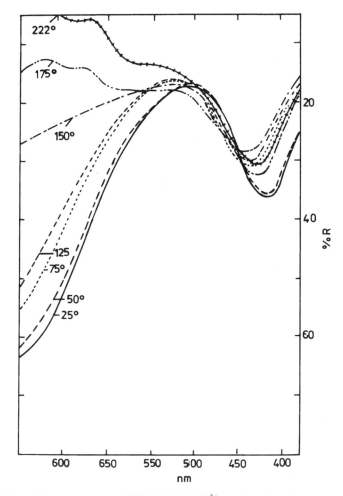

Figure 3.34 Reflectance spectra of $[Co(NH_3)_5(H_2O)]^{3+}$ as a function of temperature [114].

of $[Co(NH_3)_4(H_2O)Cl]Cl_2$, and in the third case, the total loss of water from the coordination sphere is connected with the formation of $[Co(NH_3)_4Cl_2]Cl$.

These Co(III) complexes (as well as others, which have not been mentioned here), are **irreversible thermoindicators**, as their transitions indicate rather precisely the temperatures of colour changes (Table 3.34).

Other data concerning thermoindicators are given in [115]

Apart from irreversible thermochromism, there is also reversible thermochromism. Without going into details, which can be found in monograph [25], it can be generally said that it is connected with a change of the geometrical structure and the coordination number as well as the phenomenon of monomer-dimer equilibrium. **Reversible thermoindicators** are also known [25].

The thermochromism of solutions of complex compounds is connected with all the causes that have been described for the solid phase. Of course, the solvent, water,

Table 3.34 Irreversible thermochromism in some complexes (thermoindicators) [115]

Compound	Colour		Temp./°C
	Low temp.	High Temp.	
$CoI_2 \cdot 2hmta \cdot 8H_2O$	Brown pink	Green	50
$NiBr_2 \cdot 2hmta \cdot 9H_2O$	Green	Blue	62
$Co(NCS)_2 \cdot 2py \cdot 10H_2O$	Lavender	Blue	93
$CoSiF_6$	Orange pink	Bright pink	99
$[Cr(en)_3]Cl_3$ (I)	Yellow	Red	119
$[Cr(en)_3]Cl_3$ (II)	Red	Black	270
$[Cr(en)_3](NCS)_3$ (I)	Yellow	Red	121
$[Cr(en)_3](NCS)_3$ (II)	Red	Black	252

and especially organic solvents, have additional specific effects. The essential issues were discussed in the section devoted to solvatochromism.

Much less attention has been paid to thermochromism in solution, perhaps because temperature effects on solvatochromic shifts are relatively small. For a series of $Mo(CO)_4(diimine)$ compounds, ν_{max} values change by between 40 and 320 for a 40 °C change in temperature (1 to 8 cm^{-1} K^{-1}), a change so small as to be difficult to estimate with any confidence [116]. The thermochromism of iron-diimine-cyanide complexes is similarly small; the temperature coefficients of iron(II) and iron(III) complexes have opposite signs, just as their solvatochromic shifts are in opposite directions [117].

3.3.2.4   Piezochromism

In the solid state, application of high pressure can cause phase and colour changes. Thus mercury(II) iodide, long known to change from a red tetragonal layer structure to a yellow orthorhombic modification at 127 °C, also changes colour from red to yellow (though a different yellow form) at about 10 kbar [118].

Piezochromic effects in solution are very small. Typically there is a shift in wavelength of maximum absorption of a few nm for a pressure increase of a few kilobars. Thus, for example, λ_{max} for $Fe(bipy)_2(CN)_2$ in aqueous solution shifts from 519 nm at atmospheric pressure to 516 nm at an applied pressure of 1.5 kbar. This shift, of +80 cm^{-1}, corresponds to a transition energy change of only 1.2 kJ mol^{-1}, in an overall transition energy of 230 kJ mol^{-1} [117]. For further information and discussion, the interested reader is referred to reports on piezochromism of pentacyanoferrates [103] and of molybdenum- and tungsten-diimine-oxocyanides [119], and overviews of the relation between piezochromism and solvatochromism [96, 103, 116, 117, 120].

3.3.2.5   Some aspects of the colour of transition metals in the solid phase

This problem has already been partially discussed above, in connection with thermochromism, and some aspects will be dealt with in the following chapters devoted to the colour of crystals, minerals, and glasses.

The prediction of the colour of chemical compounds in the solid phase is more difficult than in solutions, as in the former case structural considerations play an essential role. Some data concerning this issue can be found in the works of Schmitz-DuMont et al. [121–123], and recently mainly in the works of Reinen et al. [124–127].

Most data have been obtained in the studies of transition metal oxides as well as oxide matrices doped with transition elements. In this area, probably most studies have been devoted to the colour of ruby. Some aspects of this issue are discussed in Chapter 7. Here, the colour of ruby (and other oxide systems) will be considered in connection with ligand field theory and structural aspects.

It is worth noting that the interest in the colour of ruby has a long history, which is due to both a desire to explain the nature of colour and practical considerations (e.g. the possibility of manufacturing synthetic rubies).

Neuhaus [128] presented valuable results concerning this issue. His interpretation was based on the results of studies of absorption and reflectance spectra and on crystal field theory with additional consideration given to aspects of crystal chemistry.

Figure 3.35 shows the spectra of 13 samples containing various amounts of Cr^{3+} as impurities as well as the spectra of glass and Cr_2O_3 presented for comparison. On the left hand side are absorption spectra of red-coloured systems, on the right hand side, green-coloured. Table 3.35, derived from the publication cited above, gives the positions of the two main absorption spectra in the Cr^{3+} ion, the symmetry of forms, the colour, and the composition of the chromophores.

In many publications concerning the colour of compounds containing Cr^{3+} (gemstones, minerals) green and red are taken as the two characteristic colours. The author of [128] rightly concludes that a change from green to red (or the other way round) does not indicate a discontinuity of physicooptical properties, but is to do with appearance, i.e. the physiological process of vision (and colour judgment). This issue is discussed more fully by Poole [129], who also determined the chromaticity coordinates using the trichromatic colorimetry method.

Structural aspects discussed in [128] include the effect of the size of the unit cell on transition energy and hence on the colour of the examined crystals. These effects were noted by one of the authors when studying single crystals of Mg-Ag spinels with a 1% weight content of Cr_2O_3. The results are presented in Figure 3.36 and 3.37.

As can be seen, a change in the proportion of MgO and Al_2O_3 brings about a clear change of the lattice constant, which in turn causes a change of the wavelength of both the maxima occurring in the absorption spectrum of Cr^{3+}. In most general terms, it can be stated that the immediate cause of such a change is the change of the metal-ligand distance in the CrO_6 chromophore.

These issues are presented in more detail in the works by Poole [129, 130]. They concentrate on solid solutions in oxide matrices. One of the papers [130] presents data concerning the Cr content, the wavenumbers of both the electronic transitions, the B values, and the colours of a total of 70 systems. Table 3.36 contains some examples taken from that paper. The tabulated data presented in that publication indicate that the extreme values of $10Dq$, which are determined by the wavenumber of the first long-wave $^4A_{2g} \rightarrow {}^4T_{2g}(P)$ transition, fall within the range 15 823–18 450 cm^{-1}, and the second transition shows wavenumbers in the range 21 700–25 000 cm^{-1}.

It can be easily seen that most of the compounds under discussion show a green or red colour, with some also showing pink, grey pink, green yellow, or olive green colours.

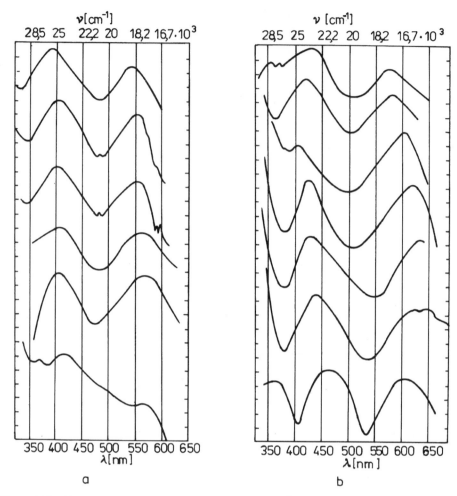

Figure 3.35 Absorption spectra of systems with different Cr^{3+} contents, as specified in brackets [128]. From the top: (a) 1 – Natural spinel (1%), 2 – synthetic ruby (pulverised; 3%), 3 – natural ruby (1%), 4 – γ-AlOOH (3%), 5 – potassium-chromium alum (single crystal; 1%), 6 – pyrope (1 to 2%); (b) 1–Chrysoberyl (1%), 2 – $MgGa_2O_4$ (spinel, pulverised; 3%), 3 – uvarovite (15%), 4 – emerald (1%), 5–fuchsite (55%), 6 – soda-phosphate glass (1 to 2%), 7 – α-Cr_2O_3 (very pure precipitated material).

Poole and Itzel [129] discussed the question of the effect of the Cr–O bond length and the angle in the O–Cr–O entity (in the CrO_6 coordination polyhedron) as well as the effect of chromium content, temperature and pressure on colour. Among other things, they found that the following dependences obtain:

$$\Delta = \Delta_1 + \Delta_2 \left(\frac{r_0}{r}\right)^5$$

$$B = B_1 + B_2 \left(\frac{r_0}{r}\right)^5,$$

Table **3.35** Absorption bands and colour of some gemstones and minerals containing Cr^{3+} [128]

No.[a]	Name (source)	Symmetry of Cr site	%Cr_2O_3	Colour	Band position[b] nm I	nm II	Structural entities[c]
1	Ruby (natural; India)	D_{3d}	<1	red	410	556	$[Al^{Cr}O_6]$
2	Ruby (synthetic; Verneuil)	D_{3d}	~1	red	410	556	$[Al^{Cr}O_6]$
3	Ruby (natural; pulverised)	D_{3d}	~1	red	405	555	$[Al^{Cr}O_6]$
5	Ruby (natural; Ceylon = Sri Lanka)	O_h	<1.5	red	390	546	$[Al^{Cr}O_6]$; $[MgO_4]$
6	$MgAl_2O_4$ (Cr-doped; synthetic)	O_h	~1	red	390	550	$[Al^{Cr}O_6]$; $[MgO_4]$
8	Potassium alum (Cr-doped; synthetic)	T_h	~0.5	red	403	565	$[Al^{Cr}(H_2O)_6]$; $[SO_4]$
13	Cr_2O_3 (spinel)	D_{3d}	100	dark green	460	600	$[CrO_6]$
14	$MgGa_2O_4$ (Cr-doped; synthetic)	O_h	3	green	420	573	$[Ga^{Cr}O_6]$; $[MgO_4]$
15	$MgCr_2O_4$	O_h	79	brown	445	575	$[CrO_6]$; $[MgO_4]$
18	Chrysoberyl	D_{2h}	~1	green	425	576	$[Al^{Cr}O_6]$; $[BeO_4]$
20	Uvarovite (Urals)	O_h	→20	dark green	403	605	$[Al^{Cr}O_6]$; $[CaO_8]$; $[SiO_4]$
22	Emerald (synthetic; Chatham)	D_{6h}	~1	green	435	603	$[Al^{Cr}O_6]$; $[BeO_4]$; $[SiO_4]$
26	Soda phosphate glass		1-2	green	440	645	$[CrO_6]$
28	$CrCl_3$			dark red	525	730	$[CrCl_6]$

[a] These numbers correspond with the numbering in ref. [128]. [b] Reflectance spectra in the case of specimens 6, 8, 13, 14, and 15.
[c] $[Al^{Cr}O_6]$ indicates isomorphous replacement of Al by Cr to the extent indicated in column 4.

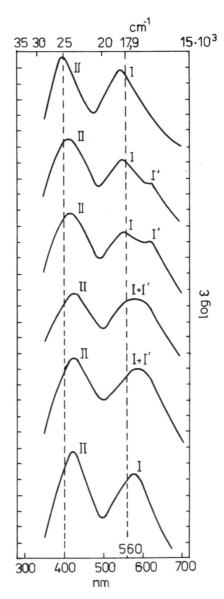

Figure 3.36 Transmission spectra of single crystals of Mg–Al spinels containing 1% by weight Cr_2O_3 for different Mg:Al ratios. From the top: 1:1 – red, 1:1.5 – red green, 1:2 – green red, 1:3 – green, 1:4 – green, 1:5 – green [128].

where r is the mean length of Cr–O, r_0 is the mean length of Al–O in pure Al_2O_3 at room temperature, Δ denotes Dq, and Δ_1, Δ_2, B_1, B_2 are constants. In the case of the solid solutions of Cr_2O_3 in Al_2O_3, the values are as follows: $\Delta_1 = 10\,600\,cm^{-1}$, $\Delta_2 = 7550\,cm^{-1}$, $B_1 = -286\,cm^{-1}$, $B_2 = 936\,cm^{-1}$.

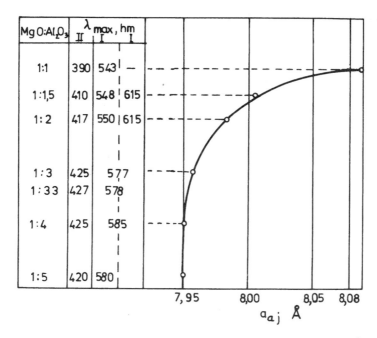

Figure 3.37 Correlation of absorption maxima from Figure 3.36 with lattice constant (a_0, Å) values [128].

Table 3.36 The dependence of $10Dq$ (Δ), T_F and B values,[a] and colour of mixed oxides $(Al_{1-x}Cr_x)_2O_3$ on the value of x [130]

No.[b]	x	$10Dq$ (Δ)	T_F	B	Colour
37	0.05	17 800	24 550	664	intense pink
38	0.15	17 600	24 100	635	dirty pink
39	0.2	17 550	23 950	623	grey pink
40	0.3	17 350	23 500	594	green grey
41	0.4	17 200	23 300	589	dirty green
42	0.6	16 950	22 700	552	dark green
43	0.8	16 700	22 100	511	dark green
44	1.0	16 600	21 700	478	dark green

[a] The symbols Δ, T_F and B (in cm^{-1}) have the same meaning as in Table 3.37. [b] The numbers of the specimens correspond to those in the original paper.

Table 3.37 presents the spectral characteristics of some systems containing Cr^{3+}.

On the basis of a series of data, it was concluded that $\Delta r^n = $ const, with n = 5 for the perovskite series of systems, and n = 2 for the ruby series (with variable chromium contents). The author postulates that in the case of different samples of ruby, chromium always occurs in the same environment and the same state. Thus, as the chromium content increases, the crystal field strength gradually decreases, and the change of colour is 'a natural consequence of this gradual change of crystal field strength' [129]. It is worth noting that the results discussed in Section 3.3.1 above,

Table 3.37 $10Dq(\Delta)$, T_F and B values,[a] and colour of some compounds containing Cr^{3+} [130]

No.[b]	Compound	x	% cont. of Cr	$10Dq$	T_F	B	Colour
1	$\alpha-(Al_{1-x}Cr_x)_2O_3$ (nat.)		<1·	17990	24390	619	red
(1)	$\alpha-(Al_{1-x}Cr_x)_2O_3$ (synth.)		0.6	17990	24390	619	red
7	$\beta-(Ga_{1-x}Cr_x)_2O_3$ (nat.)		3	16340	22624	621	green
8	$\alpha-(Ga_{1-x}Cr_x)_2O_3$ (nat.)		3	16584	22883	620	yellow green
9	$Mg(Al_{1-x}Cr_x)_2O_4$ (nat.)		<1.5	18349	25641	728	red
14	Ruby (synth.)		31.2	17857	24096	600	red
22	Muscovite (fuchsite)		4.8	15823	23584	852	green
27	$Be_3(Al_{1-x}Cr_x)_2Si_6O_{18}$ (emerald synth.)		~1	16585	22989		green
28	$Mg(Al_{1-x}Cr_x)_2O_4$	0.05		18100	~25800	787	light pink
29	$Mg(Al_{1-x}Cr_x)_2O_4$	0.1		18000	~25000	694	pink

[a] $\Delta = 10Dq$ corresponds to the $^4A_{2g} \rightarrow {}^4T_{2g}$ transition, T_F to the $^4A_{2g} \rightarrow {}^4T_{2g}$ transition (in O_h symmetry), and B is the Racah parameter; all values are in cm^{-1}. [b] The numbers of the specimens correspond to those in the original paper.

and especially the computer simulation of the spectra and colour of chromium(III) complexes as a function of Dq and the resulting changes of absorption band positions, are fully consistent with these experimental data.

Some of the results and discussion which have been presented in this Chapter have appeared previously [131]. Colour photographs of some of the compounds and complexes which appear in this Chapter, and in Chapters 5 and 7, can be found in the colour plate section.

3.3.3   Colour, kinetics, and mechanism

The colour change usually apparent on carrying out a substitution or electron transfer reaction involving one or more transition metal species can form the basis of spectrophotometric monitoring of the kinetics of such reactions. In view of the frequency with which such monitoring of reaction progress is employed, both for fast reactions and those of conventional half-lives, the phenomenon of colour underpins a great deal of our knowledge of the kinetics and mechanisms of reactions of transition metal complexes. This is not the place to discuss this very large area, whose development over the past half-century has been well chronicled, starting with the seminal book of Basolo and Pearson [132], and continuing with a number of texts [133] up to the present time [134]. Rather we shall illustrate the key role played by colour in a few aspects of electron transfer processes [135]. These examples involve the transient colours sometimes observed in certain types of electron transfer reactions, and the solvatochromism of metal-metal (or inter-valence) charge-transfer (MMCT or IVCT) bands in mixed valence complexes as a source of information on rate constants for fast intramolecular electron transfer.

An early example of an intensely coloured transient intermediate was that reported for Cr^{2+}aq reduction of $[IrCl_6]^{2-}$. The green intermediate en route to the olive-brown final solution was taken to be the binuclear species $[(H_2O)_5CrClIrCl_5]$. After electron transfer the metal centres are d^3 chromium(III) and d^6 (t_{2g}^6)

iridium(IV). As these are both substitution-inert a significant time elapses before the binuclear complex dissociates, i.e. before the blue colour fades. A later reinvestigation [136] showed this $Cr^{2+}/[IrCl_6]^{2-}$ reaction to be more complicated, with three parallel pathways, but this does not detract from the importance of the coloured intermediate in what is now believed to be the dominant, rather than the sole, electron transfer path. A similar example is provided by Cr^{2+}aq reduction of oxouranium(VI), UO_2^{2+}aq, where the dark green colour and the spectrum of the intermediate permit the assignment of oxidation states $[Cr^{III}OU^VO]^{4+}$. Again the chromium(III) is substitution-inert thanks to its d^3 electron configuration, while this time the 5+ formal charge on the uranium discourages U–O bond breaking.

Chromium(II) reduction of several cobalt(III) complexes $[Co(NH_3)_5L)]^{n+}$, where L is an easily reducible ligand such as phenylglyoxal or pyrazine-2-carboxylate, also usually gives intensely coloured intermediates [137]. These can often be shown, by ESR spectroscopy, to contain radical ligands corresponding to one-electron reduction of L. Thus the electron transfer mechanism must involve the electron dwelling on the ligand for a significant time before moving on to the cobalt centre. The reluctance of the electron to move onto the cobalt can be attributed to the barrier to transfer from a π-orbital on the ligand to the vacant e_g orbital, of σ symmetry, on the cobalt(III). This two-stage process is sometimes called the 'chemical mechanism' for electron transfer [138]. The chromium(III) complex $[Cr(H_2O)_5(phenylglyoxalate)]^{2+}$ reacts with Cr^{2+}aq to give a dark green species, best formulated as a radical-bridged binuclear complex $[Cr^{III}-L^\bullet-Cr^{III}]^{4+}$.

The intermediacy of a radical-ligand intermediate may occur in an inner-sphere or an outer-sphere redox reaction. A reaction which indubitably involves inner-sphere electron transfer is the reduction of the pyrazine carboxylate complex $[Co^{III}(en)_2(pzc)]^{2+}$ by $[Fe^{II}(CN)_5(H_2O)]^{3-}$. Here electron transfer takes place through a deep-blue radical ligand bridged intermediate, with transfer of the pyrazine carboxylate to the iron [139]. In this example the intermediate is formed rapidly on mixing the reagents, the half-life for water replacement in $[Fe^{II}(CN)_5(H_2O)]^{3-}$ being of the order of seconds or less. It is considerably more difficult to replace water bonded to cobalt(III). Reaction of trans-$[Co(en)_2(NH_3)(H_2O)]$ with $[Fe(CN)_6]^{4-}$ is accompanied by the slow development of an intense red colour, due to $Fe^{II} \rightarrow Co^{III}$ MMCT (IVCT), at a rate corresponding to substitution of the coordinated water.

Electron transfer may also occur by the outer-sphere mechanism within an ion-pair rather than in a binuclear intermediate. In such cases the ion-pair generally has an intense colour, due to intermolecular charge-transfer – again MMCT (IVCT) in character. The crystallisable violet salt $[Ru(NH_3)(py)]_4[Fe(CN)_6]_3 7H_2O$ is a particularly stable example of such a species. Diquat and paraquat salts of the type $[dq]_2[M(CN)_8]$ are examples of organic analogues; the brownish red molybdenum and green tungsten ion-pairs are formed on mixing a colourless solution of diquat dihalide and pale yellow $[M(CN)_8]^{4-}$ [140].

A complementary source of coloured transients in redox reactions is iron(III) oxidation (Fe^{3+}aq) of organic molecules such as thiourea, cysteine, organic thiols, or thiocarboxylates. Here there is generally rapid equilibrium formation of a thio-ligand-iron(III) complex, whose colour (e.g. intense blue for thiomalate) fades with time as the electron transfers relatively slowly to the iron [141].

In contrast to the reactions discussed in the preceding paragraphs, where electron transfer is on a timescale of seconds or minutes, there are many systems where an electron moves from one metal to another extremely rapidly. Thus, for example, $[IrBr_6]^{3-}$ reduction of $[Fe(4, 7\text{-}Me_2phen)_3]^{3+}$ or electron exchange in the $[Mo_4S_4\text{-}(edta)_2]^{4-/2-}$ system are characterised by rate constants as high as 10^{10} dm^3 mol^{-1} s^{-1}. Interpretation of kinetic data on outer-sphere processes such as these is straightforward; it is more difficult to deal with reactivity in inner-sphere redox reactions. In this latter class the overall redox process may be broken down into as many as six individual steps, the extra complication arising from the bond-making and breaking processes intrinsic to inner-sphere electron transfer [142]. The actual electron transfer step takes place in a binuclear species M–L–M′. This step may be slow, as in the special case of $M = Co^{III}$ discussed above. More usually it is very fast, and often very difficult to estimate, even when working with tailor-made M–L–M′ complexes.

In appropriate cases rate constants for M′ → M electron transfer in M-L-M′ complexes may be estimated from line-broadening in ESR or Mössbauer spectra, but the number of elements with suitable nuclei is limited, in the present context essentially to copper for the former, iron for the latter. However, from a preparative and kinetic point of view, the most suitable metal for the study of electron transfer is often ruthenium. This element has a variety of coloured mixed valence bi- and tri-nuclear compounds, such as ruthenium red, ruthenium brown, ruthenium blues, and the Creutz-Taube cation $[(H_3N)_5Ru(\mu\text{-pyrazine})Ru(NH_3)_5]^{5+}$. The great advantage of ruthenium mixed valence complexes is that both ruthenium(II) and ruthenium(III) are substitution-inert. In such complexes it is sometimes possible to estimate rate constants for electron transfer from metal-to-metal charge-transfer (MMCT) bands [143], though there are two restrictions. The first is that the MMCT must not be obscured by the more intense MLCT (or LMCT) bands usually also present. As MMCT bands are usually towards, or just in, the infrared region, they are in practice often sufficiently separated from the MLCT bands to be observed and examined. The second restriction is that the mixed valence compound must be in Class II of the Robin and Day classification [144]. Such compounds have the metal ions in the two oxidation states in crystallographically distinguishable sites, but there is significant electron delocalisation. In other words the two metals must be in limited communication – neither with localised oxidation states and a very high barrier to electron transfer (Class I, as in e.g. Pb_3O_4 or $[Co(NH_3)_6]_2[CoCl_4]_3$) nor with complete delocalisation (Class III, as in e.g. $[Nb_6Cl_{12}]^{2+}$ or bioinorganic Fe_4S_4 units). Class II compounds include such intensely coloured species as Eu_3S_4 (black, with a reddish tinge), Prussian Blue, and some platinum(II)-platinum(IV) chain compounds. It is the solvatochromic behaviour of the observed charge-transfer band(s) which is diagnostic of Class II behaviour.

Once appropriate behaviour has been established, the calculations and assumptions involved on going from colour and spectrum through to kinetics are as follows. The first step is to calculate the barrier to electron transfer. Next it has to be assumed, quite reasonably, that electron transfer is adiabatic, and a value has to be assumed for the transmission coefficient. It is then possible to calculate a rate constant from the energy barrier, though implicit in this calculation is that the activation entropy is close to zero. This depends on solvation changes consequent on electron transfer being small – probably a reasonable assumption for mixed valence species of the $[L_5M-LL-ML'_5]^{5+}$ type {e.g. ruthenium(II)/(III)-ammines} but unlikely to hold for e.g. $[(H_3N)_5Ru-LL-Fe(CN)_5]$ in view of the known large differences in hydration of $[Fe(CN)_6]^{3-}$ and $[Fe(CN)_6]^{4-}$. Nonetheless, despite all these estimates and assumptions it is possible to obtain reasonable estimates for rate constants for electron transfer, for example the value of $3 \times 10^8 \, s^{-1}$ for $[(H_3N)_5Ru-(\mu\text{-pyrazine})-Ru(NH_3)_5]^{5+}$. In view of the central role played by mixed valence diruthenium species in intramolecular electron transfer kinetics, this application of colour and solvatochromism is of considerable importance in understanding reactivity and mechanisms in redox chemistry of transition metal complexes, and indeed also of a number of organo-transition metal compounds [145].

REFERENCES

[1] Frémy, E., 1852, *Ann. Chim. (Paris)* [3], **35**, 257; *J. Prakt. Chem.*, **57**, 95.

[2] Sidgwick, N.V., 1950, *The Chemical Elements and Their Compounds, Vol. 2*, Clarendon Press, Oxford, pp. 1413–5; Kauffmann, G.B., 1981, *Inorganic Coordination Compounds*, Heyden, London, pp. 5, 8–9, and Chapter 6.

[3] Ballhausen, C.J., 1962, *Introduction to Ligand Field Theory*, McGraw-Hill, New York.

[4] Bartecki, A., 1971, *Spektroskopia elektronowa związków nieorganicznych i kompleksowych*, PWN, Warsaw.

[5] Figgis, B.N., 1966, *Introduction to Ligand Fields*, Interscience, New York.

[6] Gołębiewski, A., 1969, *Chemia kwantona związków nieorganicznych*, PWN, Warsaw.

[7] Jørgensen, C.K., 1962, *Absorption Spectra and Chemical Bonding in Complexes*, Academic Press, London.

[8] Lever, A.B.P., 1984, *Inorganic Electronic Spectroscopy*, 2nd edn., Elsevier, Amsterdam.

[9] Bartecki, A., Myrczek, J., Staszak, Z., Waśko, K., Sowińska, M. and Kurzak, K., 1987, *Widma elektronowe związków kompleksowych. Metody analizy, przetwarzania i gromadzenia przy uzyciu mikrokomputera*. WNT, Warsaw.

[10] Fajans, K., 1923, *Naturwissenschaften*, **11**, 165.

[11] Tsuchida, R., 1938, *Bull. Chem. Soc. Japan*, **13**, 388, 434, 471.

[12] Jørgensen, C.K., 1962, *Absorption Spectra and Chemical Bonding in Complexes*, Pergamon Press, Oxford, Chapter 7.

[13] Jørgensen, C.K., 1971, *Modern Aspects of Ligand Field Theory*, North Holland, Amsterdam.

[14] Beck, M.T. and Porszolt, É.C., 1971, *J. Coord. Chem.*, **1**, 57; Beck, M.T., 1972, in *Coordination Chemistry in Solution*, ed. Högfeldt, E., Swedish National Science Research Council, Lund, p. 241.

[15] See, e.g., Pfeil, A., Palmer, D.A. and Kelm, H., 1980, *Spectrochim. Acta*, **36A**, 1013.

[16] Cotton, F.A., Goodgame, D.M.L., Goodgame, M. and Secco, A., 1961, *J. Am. Chem. Soc.*, **83**, 4157.

[17] Condon, E.U. and Shortley, G.H., 1962, *The Theory of Atomic Spectra*, Cambridge University Press.

[18] Racah, G., 1949, *Phys. Rev.*, **76**, 1352.

[19] Jørgensen, C.K., 1970, *Progr. Inorg. Chem.*, **12**, 159.

[20] Schäffer, C.E. and Jørgensen, C.K., 1958, *J. Inorg. Nucl. Chem.*, **8**, 143

[21] Pauling, L., 1948, *The Nature of the Chemical Bond*, 2nd edn., Cornell University Press, Ithaca, New York.

[22] Kurzak, K. and Bartecki, A., 1988, *Transition Met. Chem.*, **13**, 224; Kurzak, K., 1991, *Spektrochemiczne własnosci niskosymetrycznych mieszanych kompleksów niklu(II), chromu(III) i miedzi(II) w roztworach*, Wyższa Szkoła Rolniczko-Pedagogiczna, Siedlce.

[23] Libuś, W., 1959, *Roczniki Chem.*, **33**, 931, 951.

[24] Starosta, J. and Bartecki, A., 1970, *Prace Naukowl Inst. Chemii Nieorg. i Met. Pierwiastków Rzadkich, Studia i Materiały*, Nr. 2, **4**, 25.

[25] Sone, K. and Fukuda, Y., 1987, *Inorganic Thermochromism*, Springer-Verlag, Berlin.

[26] Bartecki, A. and Myrczek, J., unpublished data (enquiries to the Institute of Inorganic Chemistry and Metallurgy of Rare Elements, Technical University of Wrocław).

[27] Jørgensen, C.K., 1957, *Acta Chem. Scand.*, **11**, 399.

[28] Sone, K. and Kato, M., 1959, *Z. Anorg. Allg. Chem.*, **301**, 277.

[29] Bartecki, A., Staszak, M. and Raczko, M., 1988, *Mater. Sci.*, **14**, 59.

[30] Jørgensen, C.K., 1963, *Orbitals in Atoms and Molecules*, Academic Press, London.

[31] Wolfsberg, M. and Helmholz, L., 1952; *J. Chem. Phys.*, **20**, 837; Helmholz, L., Brennan, H. and Wolfsberg, M., 1955, *J. Chem. Phys.*, **23**, 853.

[32] Ballhausen, C.J. and Liehr, A.D., 1958, *J. Mol. Spectroscopy*, **2**, 342.

[33] Viste, A. and Gray, H.B., 1964, *Inorg. Chem.*, **3**, 1113.

[34] Miller, R.M., Tinti, D.S. and Case, D.A., 1989, *Inorg. Chem.*, **28**, 2738.

[35] Carrington, A., Schonland, D. and Symons, M.C.R., 1957, *J. Chem. Soc.*, 659.

[36] Teltow, J., 1939, *Z. Phys. Chem.*, **B43**, 198.

[37] Bartecki, A., 1970, *Ionic Radii and Electronic Transition Energies in d^0 Coordination Compounds*, Section Lecture, XIII ICCC, Kraków-Zakopane.

[38] Williams, A.F., 1986, *Chemia nieorganiczna. Podstawy teoretyczne*, PWN, Warsaw, p. 208; 1979, *A Theoretical Approach to Inorganic Chemistry*, Springer Verlag, Berlin.

[39] Tłaczała, T., Cieślak-Golonka, M., Bartecki, A. and Raczko, M., 1993, *Applied Spectroscopy*, **47**, 1704.

[40] Cieślak-Golonka, M. and Bartecki, A., 1978, *Bull. Acad. Polon. Sci. Ser. Sci. Chem.*, **26**, 53.

[41] Cieślak-Golonka, M. and Bartecki, A., 1979, *Pol. J. Chem.*, **53**, 743.

[42] Bartecki, A. and Dembicka, D., 1973, *Inorg. Chim. Acta*, **7**, 610.

[43] E.g., Bartecki, A., Ph.D. Thesis, 1960; Jeżowska-Trzebiatowska, B. and Bartecki, A., 1962, *Spectrochim. Acta*, **18**, 799; Bartecki, A. and Myrczek, J., 1984, *Computer Enhanced Spectroscopy*, **2**, 53.

[44] Bartecki, A., 1964, *Roczniki Chem.*, **38**, 1455.

[45] Bartecki, A. and Dembicka, D., 1965, *Roczniki Chem.*, **39**, 1793.
[46] Bartecki, A. and Dembicka, D., 1967, *J. Inorg. Nucl. Chem.*, **29**, 2907.
[47] Flaschka, H., *Talanta*, **7**, 90 (1960).
[48] Reilley, C.N., Flaschka, H.A., Laurent, S. and Laurent, B., 1960, *Analyt. Chem.*, **32**, 1218.
[49] Myrczek, J., unpublished data (enquiries to the Institute of Inorganic Chemistry and Metallurgy of Rare Elements, Technical University of Wrocław).
[50] Greenwood N.N. and Earnshaw, A., 1998, *Chemistry of the Elements*, 2nd edn., Pergamon, Oxford.
[51] Volkov, K. and Yatsimirskii, K.B., 1977, *Naukowa Dumka, Kiev*, p. 66.
[52] Luo, M.R. and Rigg, B., 1987, *J. Soc. Dyers Colourists*, **103**, 86, 126.
[53] Bartecki, A. and Staszak, Z., 1988, *Bull. Pol. Acad. Chem.*, **36**, 373.
[54] Zausznica, A., 1989, *Nauka o barwie*, PWN, Warsaw.
[55] Bartecki, A. and Tłaczała, T., 1990, *Spectr. Lett.*, **23**, 727.
[56] Griffiths, J., 1976, *Colour and Constitution of Organic Molecules*, Academic Press, London.
[57] Reichardt, C., 1968, *Lösungsmitteleffekte in der Organischen Chemie*, Verlag Chemie, Weinheim; Reichardt, C., 1988, *Solvents and Solvent Effects in Organic Chemistry*, 2nd edn., VCH, Weinheim; Suppan, P. and Ghoneim, N., 1997, *Solvatochromism*, RSC, Cambridge.
[58] Spange, S., Hortschansky, P., Ulbricht, A. and Heublein, G., 1987, *Z. Chem.*, **27**, 207.
[59] Spange, S., Hortschansky, P. and Heublein, G. 1989, *Acta Polymerica*, **40**, 602.
[60] Dutta, P.K. and Turbeville, W., 1991, *J. Phys. Chem.*, **95**, 4087.
[61] See, e.g., Deye, J.F., Berger, T.A. and Anderson, A.G., 1990, *Analyt. Chem.*, **62**, 615.
[62] Razak bin Ali and Burgess, J., 1993, *Transition Met. Chem.*, **18**, 9; and references cited therein.
[63] Bartecki, A., Tłaczała, T. and Raczko, M., 1991, *Spectr. Lett.*, **24**, 559.
[64] Bartecki, A. and Tłaczała, T., 1993, *Spectr. Lett.*, **26**, 809.
[65] Böttcher, C.J.F., 1962, *Theory of Electric Polarization*, Elsevier, Amsterdam.
[66] Onsager, L., 1936, *J. Am. Chem. Soc.*, **58**, 1486.
[67] Bartecki, A. and Stelmaszek, L., 1981, *Bull. Pol. Acad. Chem.*, **29**, 307.
[68] Myrczek, J. and Bartecki, A., 1987, *Spectr. Lett.*, **20**, 431.
[69] Shibuya, T., 1983, *J. Chem. Phys.*, **78**, 5175.
[70] Shibuya, T., 1984, *Bull. Chem. Soc. Japan*, **57**, 2991.
[71] Staszak, Z. and Bartecki, A., 1989, *Spectr. Lett.*, **22**, 1193.
[72] Gutmann, V., 1971, *Chemische Funktionslehre*, Springer-Verlag, Vienna.
[73] Griffiths, T.R. and Pugh, D.C., 1979, *Coord. Chem. Rev.*, **29**, 129.
[74] Fujiwara, M., Yoshitake, M., Fukuda, Y. and Sone, K., 1988, *Bull. Chem. Soc. Japan*, **61**, 2967.
[75] Fukuda, Y. and Sone, K., 1972, *Bull. Chem. Soc. Japan*, **45**, 465; Scremin, M., Zanotto, S.P., Machado, V.G. and Rezende, M.C., 1994, *J. Chem. Soc., Faraday Trans.*, **90**, 865.
[76] Migron, Y. and Marcus, Y., 1991, *J. Phys. Org. Chem.*, **4**, 310.
[77] Barbieri, G.A., 1934, *Atti Accad. Nazl. Lincei*, **20**, 273.

[78] Schilt, A.A., 1960, *J. Am. Chem. Soc.*, **82**, 3000, 5779; **85**, 904 (1963).

[79] Dimroth, K., Reichardt, C., Siepmann, T. and Bohlmann, F., 1963, *Liebigs Ann. Chem.*, **661**, 1.

[80] Reichardt, C., 1965, *Angew. Chem. Int. Edn. Engl.*, **4**, 29; Reichardt, C., 1979, *Angew. Chem. Int. Edn. Engl.*, **18**, 98; Reichardt, C. and Harbusch-Görnert, E., 1983, *Liebigs Ann. Chem.*, 721.

[81] Soukup, R.W., 1983, *Chemie in unserer Zeit*, **17**, 129.

[82] Soukup, R.W. and Schmid, R., 1985, *J. Chem. Educ.*, **62**, 459.

[83] See, e.g., Blandamer, M.J., Burgess, J. and Haines, R.I., 1981, *Transition Met. Chem.*, **6**, 145.

[84] Bjerrum, J., Adamson, A.W. and Bostrup, O., 1956, *Acta Chem. Scand.*, **10**, 329.

[85] Spange, S., Lauterbach, M., Gyra, A.-K. and Reichardt, C., 1991, *Liebigs Ann. Chem.*, 323.

[86] Al-Alousy, A. and Burgess, J., 1990, *Inorg. Chim. Acta*, **169**, 167; Spange, S. and Keutel, D., 1992, *Liebigs Ann. Chem.*, 423; Podsiadla, M., Rzeszotarska, J. and Kalinowski, M.K., 1994, *Monatsh. Chem.*, **125**, 827.

[87] Shraydeh, B.F. and Burgess, J., 1993, *Monatsh. Chem.*, **124**, 877.

[88] Burgess, J., Radulović, S. and Sanchez, F., 1987, *Transition Met. Chem.*, **12**, 529.

[89] Burgess, J. and Morton, S.F.N., 1972, *J. Chem. Soc., Dalton Trans.*, 1712.

[90] Podsiadla, M., Rzeszotarska, J. and Kalinowski, M.K., 1994, *Coll. Czech. Chem. Comm.*, **59**, 1349.

[91] Spange, S., Gürs, K.-H., Heublein, G. and Klemm, E., 1984, *Z. Chem.*, **24**, 154; Spange, S. and Heublein, G., 1985, *Z. Chem.*, **25**, 288; Rodrigues, C.A., Stadler, E. and Rezende, M.C., 1991, *J. Chem. Soc., Faraday Trans.*, **87**, 701.

[92] Blandamer, M.J., Burgess, J. and Shraydeh, B., 1993, *J. Chem. Soc., Faraday Trans.*, **89**, 531; Burgess, J., Patel, M.S. and Tindall, C., 1993, *Spectr. Lett.*, **26**, 1469; Burgess, J., Lane, R.C., Singh, K., de Castro, B. and Gameiro dos Santos, A.P., 1994, *J. Chem. Soc., Faraday Trans.*, **90**, 3071.

[93] Sinn, H.A., Ph.D. Thesis, Darmstadt, 1966; Saito, H., Fujita, J. and Saito, K., 1968, *Bull. Chem. Soc. Japan*, **41**, 863; Overton, C. and Connor, J.A., 1982, *Polyhedron*, **1**, 53; Dodsworth, E.S. and Lever, A.B.P., 1990, *Coord. Chem. Rev.*, **97**, 271; Stufkens, D.J., 1990, *Coord. Chem. Rev.*, **104**, 39.

[94] Demas, J.N., Turner, T.F. and Crosby, G.A., 1969, *Inorg. Chem.*, **8**, 674; Belser, P., von Zelewsky, A., Juris, A., Barigelletti, F. and Balzani, V., 1985, *Gazz. Chim. Ital.*, **115**, 723.

[95] Elias, H., Macholdt, H.-T., Wannowius, K.J., Blandamer, M.J., Burgess, J. and Clark, B., 1986, *Inorg. Chem.*, **25**, 3048; Banerjee, P. and Burgess, J., 1988, *Inorg. Chim. Acta*, **146**, 227.

[96] Razak bin Ali, Burgess, J. and Guardado, P., 1988, *Transition Met. Chem.*, **13**, 126.

[97] Wrighton, M. and Morse, D.L., 1974, *J. Amer. Chem. Soc.*, **96**, 998.

[98] Al-Alousy, A., Burgess, J., Samotus, A. and Szklarzewicz, J., 1991, *Spectrochim. Acta.* **47A**, 985.

[99] Blandamer, M.J., Burgess, J. and Haines, R.I., 1976, *J. Chem. Soc., Dalton Trans.* 1293.

[100] Shepherd, R.E., Hoq, M.F., Hoblack, N. and Johnson, C.R., 1984, *Inorg. Chem.*, **23**, 3249.

[101] Warner, L.W., Hoq, M.F., Myser, T.K., Henderson, W.W. and Shepherd, R.E., 1986, *Inorg. Chem.*, **25**, 1911; Moyá, M.L., Rodriguez, A. and Sánchez, F., 1991, *Inorg. Chim. Acta*, **188**, 185.

[102] Rodríguez, A., Sánchez, F., Moyá, M.L., Burgess, J., Al-Alousy, A., 1991, *Transition Met. Chem.*, **16**, 445.

[103] Alshehri, S., Burgess, J., Morgan, G.H., Patel, B. and Patel, M.S., 1993, *Transition Met. Chem.*, **18**, 619.

[104] Burgess, J. and Patel, M.S., 1993, *J. Chem. Soc., Faraday Trans.*, **89**, 783.

[105] Sakata, K., Nakamura, H. and Hashimoto, M., 1984, *Inorg. Chim. Acta*, **83**, L67.

[106] Burgess, J., Davies, D.L., Grist, A.J., Hall, J.A. and Parsons, S.A., 1994, *Monatsh. Chem.*, **125**, 515.

[107] Burgess, J., Fawcett, J., Haines, R.I., Russell, D.R. and Singh, K., submitted to *Transition Met. Chem.*

[108] Rodríguez, A., Munoz, E., Sánchez, F., Moyá, M.L. and Burgess, J., 1992, *Transition Met. Chem.*, **17**, 5.

[109] Taura, T., 1990, *Bull. Chem. Soc. Japan*, **63**, 1105.

[110] Kaizaki, S. and Takemoto, H., 1980, *Inorg. Chem.*, **29**, 4960.

[111] Dunn, T.M., 1960, in *Modern Coordination Chemistry*, Lewis, J. and Wilkins, R.G. (eds.) New York: Interscience.

[112] Wilke, K.T. and Opfermann, W., 1963, *Z. Phys. Chem. (Leipzig)*, **224**, 237.

[113] Nagase, K., 1978, *Thermochim. Acta*, **23**, 283.

[114] Wendlandt, W. and Smith, J.P., 1967, *The Thermal Properties of Transition Metal Ammine Complexes*, Elsevier, Amsterdam.

[115] Cowling, J.E., King, P. and Alexander, A.L., 1953, *Ind. Eng. Chem.*, **45**, 2317

[116] Razak bin Ali, Burgess, J., Kotowski, M. and van Eldik, R., 1987, *Transition Met. Chem.*, **12**, 230.

[117] Kotowski, M., van Eldik, R., Razak bin Ali, Burgess, J. and Radulović, S., 1987, *Inorg. Chim. Acta*, **131**, 225

[118] Mikler, H., 1973, *Monatsh. Chem.*, **103**, 110; Adams, D.M. and Appleby, R., 1976, *J. Chem. Soc., Chem. Commun.*, 975; Adams, D.M. and Appleby, R., 1978, *Inorg. Chim. Acta*, **26**, L43..

[119] Nowicka, B., Burgess, J., Parsons, S.A., Samotus, A. and Szklarzewicz, J., 1998, *Transition Met. Chem.*, **23**, 317.

[120] Razak bin Ali, Banerjee, P., Burgess, J. and Smith, A.E., 1988, *Transition Met. Chem.*, **13**, 107; Burgess, J., 1989, *Spectrochim. Acta*, **45A**, 159; Burgess, J., Maguire, S., McGranaghan, A., Parsons, S.A., Nowicka, B. and Samotus, A., 1998, *Transition Met. Chem.*, **23**, 615.

[121] Schmitz-DuMont, O., Brokopf, H. and Gössling, H., 1959, *Z. Anorg. Allg. Chem.*, **300**, 159.

[122] Schmitz-DuMont, O., Lulé, A. and Reinen, D., 1965, *Ber. Bunsenges. Phys. Chem.*, **69**, 76.

[123] Schmitz-DuMont, O. and Grimm, D., 1965, *Monatsh.*, **96**, 922.

[124] Lachwa, H. and Reinen, D., 1989, *Inorg. Chem.*, **28**, 1044.

[125] Propach, V., Reinen, D., Drenkhahn, H. and Müller-Buschbaum, H.K., 1978, *Z. Naturforsch.*, **32b**, 619.

[126] Reinen, D., 1965, *Ber. Bunsenges. Phys. Chem.*, **69**, 82.

[127] Reinen, D., 1966, *Theor. Chim. Acta*, **5**, 312.

[128] Neuhaus, A., 1960, *Z. Kryst.*, **113**, 195

[129] Poole, C.P. and Itzel, J.F., 1963, *J. Chem. Phys.*, **39**, 3445.

[130] Poole, C.P., 1964, *J. Phys. Chem. Solids*, **25**, 1169.

[131] Bartecki, A., 1992, *Rev. Inorg. Chem.*, **12**, 35.

[132] Basolo, F. and Pearson, R.G., 1957, *Mechanisms of Inorganic Reactions – A Study of Metal Complexes in Solution*, Wiley, New York; 1968, 2nd edn, Wiley, New York.

[133] E.g., Tobe, M.L., 1972, *Inorganic Reaction Mechanisms*, Nelson, London; Wilkins, R.G., 1974, *The Study of Kinetics and Mechanism of Reactions of Transition Metal Complexes*, Allyn and Bacon, Boston; Atwood, J.D., 1985, *Inorganic and Organometallic Reaction Mechanisms*, Brooks/Cole, Monterey; Wilkins, R.G., 1991, *Kinetics and Mechanism of Reactions of Transition Metal Complexes*, VCH, Weinheim.

[134] Tobe, M.L. and Burgess, J., 1999, *Inorganic Reaction Mechanisms*, Addison-Wesley-Longman, Harlow.

[135] Cannon, R.D., 1980, *Electron Transfer Reactions*, Butterworths, London; Lappin, A.G., 1994, *Redox Mechanisms in Inorganic Chemistry*, Ellis Horwood, Chichester.

[136] Taube, H. and Myers, H., 1954, *J. Am. Chem. Soc.*, 76, 2103; Thorneley, R.N.F. and Sykes, A.G., 1970, *J. Chem. Soc. (A)*, 232; Melvin, W.S. and Haim, A., 1977, *Inorg. Chem.*, **16**, 2016.

[137] Spiecker, H. and Wieghardt, K., 1977, *Inorg. Chem.*, **16**, 1290; Boucher, H.A., Lawrance, G.A., Sargeson, A.M. and Sangster, D.F., 1983, *Inorg. Chem.*, **22**, 3482; Holaway, W.F., Srinivasan, V.S. and Gould, E.S., 1984, *Inorg. Chem.*, **23**, 2181.

[138] Haim, A., 1975, *Accts. Chem. Res.*, **8**, 264; Gould, E.S., 1985, *Accts. Chem. Res.*, **18**, 22.

[139] Malin, J.M., Ryan, D.A. and O'Halloran, T.V., 1978, *J. Am. Chem. Soc.*, **100**, 2097; Razak bin Ali, Blandamer, M.J., Burgess, J., Guardado, P. and Sanchez, F., 1987, *Inorg. Chim. Acta*, **131**, 59; van Eldik, R. and Guardado, P., 1990, *Inorg. Chem.*, **29**, 3477.

[140] Nowicka, B. and Samotus, A., 1998, *J. Chem. Soc., Dalton Trans.*, 1021.

[141] See, e.g., Ellis, K.J. and McAuley, A., 1973, *J. Chem. Soc., Dalton Trans.*, 1533; Pustelnik, N. and Sołoniewicz, R., 1975, *Monatsh. Chem.*, **106**, 673; Burgess, J., 1978, *Metal Ions in Solution*, Ellis Horwood, Chichester, Chapter 13.6.1; Table 9.38 of reference 134.

[142] See Chapter 9.4 of reference 134.

[143] Meyer, T.J., 1979, *Chem. Phys. Lett.*, **64**, 417; Yeh, A. and Haim, A., 1985, *J. Am. Chem. Soc.*, **107**, 369.

[144] Robin, M.B. and Day, P., 1967, *Adv. Inorg. Chem. Radiochem.*, **10**, 247; Clark, R.J.H., 1984, *Chem. Soc. Rev.*, **13**, 219.

[145] Burgess, J. and Pelizzetti, E., 1992, *Progr. React. Kinet.*, **17**, 1.

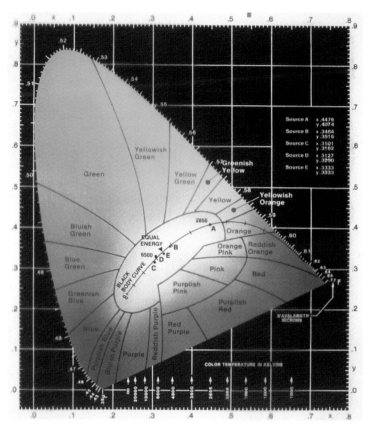

Figure C1 CIE 1931 chromaticity diagram, including points for the standard illuminants (sources) A, B, C, D_{65}, and E.

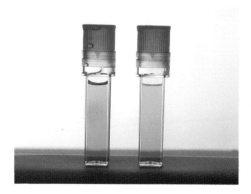

Figure C2 Aqueous solutions containing the complexes $[Fe(phen)_3]^{2+}$ (left) and $[Fe(5NO_2phen)_3]^{2+}$ (right). The visible absorption spectra of these two complexes are given in Figure 2.1.

Figure C3 Solvatochromism: Solutions of $[Fe(Schiff\ base)_2(CN)_2]$ {Schiff base from 2-acetyl-pyridine and ammonia} in 1 – water; 2 – methanol; 3 – iso-propanol; 4 – dimethyl sulphoxide (λ_{max} = 560, 590, 604, 640 nm respectively).

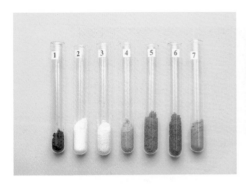

Figure C4 1 – VOSO$_4$.5H$_2$O;
2 – MnSO$_4$.7H$_2$O; 3 – FeSO$_4$.7H$_2$O;
4 – CoSO$_4$.7H$_2$O; 5 – NiSO$_4$.7H$_2$O;
6 – CuSO$_4$.5H$_2$O; 7 – Cr$_2$(SO$_4$)$_3$.18H$_2$O.

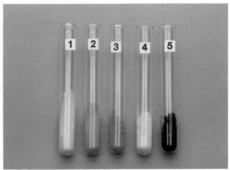

Figure C5 1 – K$_2$CrO$_4$; 2 – K$_2$Cr$_2$O$_7$; 3 – V$_2$O$_5$;
4 – WO$_3$; 5 – KMnO$_4$.

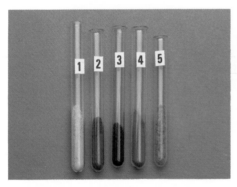

Figure C6 1 – PbCrO$_4$; 2 – CuCrO$_4$;
3 – [Ag(bipy)$_2$]$_2$CrO$_4$; 4 – [Cu(bipy)$_2$]Cr$_2$O$_7$;
5 – NH$_4$[CrO$_3$F].

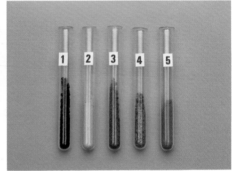

Figure C7 1 – K[Cr(C$_2$O$_4$)$_3$].2H$_2$O;
2 – [Cr(en)$_3$]$_2$(SO$_4$)$_3$; 3 – K$_3$[Cr(NCS)]$_6$.4H$_2$O;
4 – [Cr(urea)$_6$]Cl$_3$; 5 – [Cr(acac)$_3$].

Figure C8 1 – Ni(py)$_4$I$_2$; 2 – Ni(py)$_4$Cl$_2$;
3 – Ni(py)$_4$(NCS)$_2$; 4 – Ni(en)$_2$(NCS)$_2$;
5 – Ni(en)$_2$(NO$_2$)$_2$.

Figure C9 Solutions of CoCl$_2$ in organic solvents:
1 – acetone; 2 – methanol; 3 – acetonitrile;
4 – formamide; 5 – dimethyl sulphoxide.

Figure C10 Aqueous CoCl$_2$ solutions at concentrations: 1 – 0.03 M; 2 – 0.07 M; 3 – 0.2 M; 4 – 0.3 M; 5 – 0.5 M.

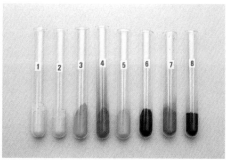

Figure C11 1 – zinc yellow; 2 – chrome yellow; 3 – chrome orange; 4 – chrome green; 5 – iron yellow; 6 – iron blue; 7 – iron red; 8 – iron black.

Figure C12   Chrysocolla

Figure C13   Wulfenite

Figure C14   Rhodochrosite

Figure C15   Malachite

4. COLOUR OF LANTHANIDE IONS AND CHEMICAL COMPOUNDS

4.1 SOME BASIC ISSUES

First of all, let us note certain points of terminology. The term 'lanthanides' is often used with reference to the group of 15 elements comprising lanthanum (La, $Z = 57$) and fourteen other elements, $Z = 58–71$, that is from cerium (Ce) to lutetium (Lu). For historical reasons, as well as because of some physicochemical properties, these elements plus scandium and yttrium are grouped under the name 'rare-earth elements', which thus comprises 17 elements of subgroup III of the periodic system.

If one considers the electronic configurations of the atoms in the ground state, given in Table 4.1, and particularly for the most characteristic chemical compounds of these elements in the +3 oxidation state, the terms *f*-**electronic elements** or **inner-transition elements** are used. The latter name means that in comparison with the **transition (*d*-electronic) elements**, the *f*-electronic configuration shows a stronger screening, resulting in more stable properties of chemical compounds. One of such properties is the colour of 4*f*-electronic element ions and compounds. The colour is the subject of the discussion below, concerning only the fourteen elements, from cerium (Ce) to lutetium (Lu), henceforth called the lanthanides.

Table 4.1 Outer electronic configurations of the lanthanides in their ground states

Ce	$4f^1 5d^1 6s^2$	Tb	$4f^9 6s^2$
Pr	$4f^3 6s^2$	Dy	$4f^{10} 6s^2$
Nd	$4f^4 6s^2$	Ho	$4f^{11} 6s^2$
Pm	$4f^5 6s^2$	Er	$4f^{12} 6s^2$
Sm	$4f^6 6s^2$	Tm	$4f^{13} 6s^2$
Eu	$4f^7 6s^2$	Yb	$4f^{14} 6s^2$
Gd	$4f^7 5d^1 6s^2$	Lu	$4f^{14} 5d^1 6s^2$

117

4.2   SPECTROSCOPIC PROPERTIES AND COLOUR OF LANTHANIDE CHEMICAL COMPOUNDS IN OXIDATION STATE +3

As has been said, the most important and stable oxidation state in lanthanides is the +3 oxidation state. The f^q equivalent electron configuration causes the occurrence of the appropriate terms and energy levels. They are given in Table 4.2, which shows that in comparison with the d^q equivalent electron configuration, here there are many more energy levels. A conclusion can be drawn that the number of electronic transitions will be much greater for elemental (gaseous) state ions, as well as for chemical compounds with these elements. The transitions occur within the f^q electronic configuration and are referred to as f-f transitions, by analogy to d-d transitions. They are forbidden by Laporte's rule, which says that transitions are allowed if the orbital quantum number l changes by one unit, thus forbidding transitions where $\Delta l = 0$.

Table 4.2 Numbers of terms and levels of the electronic configurations f^q in Ln^{3+} ions[a]

		number of	
		terms	levels
f^1, f^{13}	Ce^{3+}, Yb^{3+}	1	2
f^2, f^{12}	Pr^{3+}, Tm^{3+}	7	13
f^3, f^{11}	Nd^{3+}, Er^{3+}	17	41
f^4, f^{10}	Pm^{3+}, Ho^{3+}	47	107
f^5, f^9	Sm^{3+}, Dy^{3+}	73	198
f^6, f^8	Eu^{3+}, Tb^{3+}	119	295
f^7	Gd^{3+}	119	327

[a] The energy levels have been clearly summarised diagrammatically in Figure 2.9 of S.A. Cotton, *Lanthanides and Actinides*, Macmillan, London, 1991 (p. 29); see S. Hüfner, in *Systematics and Properties of the Lanthanides*, ed. S.P. Sinha, Reidel, Dordrecht, 1983, for further information. Ground states for the Ln^{3+} ions are given in Table 4.3.

Electronic transitions in the 380–780 nm range, which are the cause of colour in chemical compounds, in the case of d-electronic elements are transitions between energy levels split in the crystal field (their source are ligands bonded to the metal ion). As is well known, the splitting depends on the symmetry of the field. In contrast to the transition elements, in the case of the lanthanides, there is a weak crystal field, weaker than the decisive interaction, i.e. the spin-orbit coupling. In accordance with the formula $L + S = J$, where L, S, and J are the quantum numbers: orbital, spin, and total (internal), the J quantum number causes the occurrence of the levels $L + S$, $L + S - 1, \ldots, |L - S|$. Thus, for instance, the term ^3P, which in d-electronic complexes does not split in a crystal field of octahedral symmetry, may undergo a splitting in the case under consideration into three levels: $^3P_0, {}^3P_1, {}^3P_2$, for $J = 0, 1, 2$.

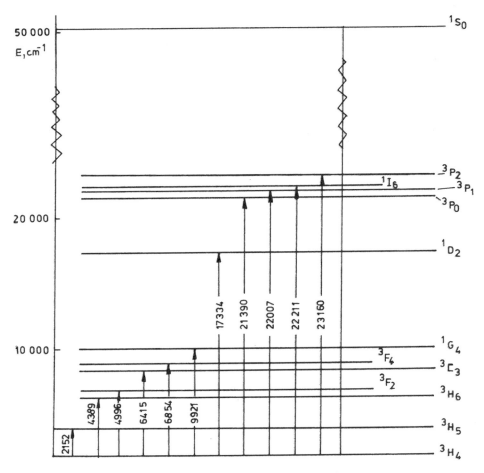

Figure 4.1   Energy levels of the free (gaseous) Pr^{3+} ion [2].

As a result of this, lanthanide ions and compounds show mostly transitions between multiplet levels, both in the ground state and in the excited states.

Examples of energy diagrams (for the Pr^{3+} gaseous ion and the Nd^{3+} aquaion) are given in Figure 4.1 and 4.2 (after [1, 2]). The diagrams illustrate the complex structure of the electronic transitions and the growing number of such transitions as the number of equivalent electrons, f, increases.

Figure 4.3 [3] shows the spectra of the aquaions of twelve Ln^{3+} lanthanides.

An important characteristic of lanthanide electronic spectra, whether absorption or emission, is that their half-width is in the order of several hundred cm^{-1} (in solution), many times less than in d-electronic element ions (compounds). The spectra can be regarded as line spectra, and they can really be obtained only by means of high resolution spectrographs (or for single crystals at liquid helium temperature).

Even though in spectrophotometric studies, especially for solutions, such spectra are not obtained, the spectra of lanthanide compounds always show this

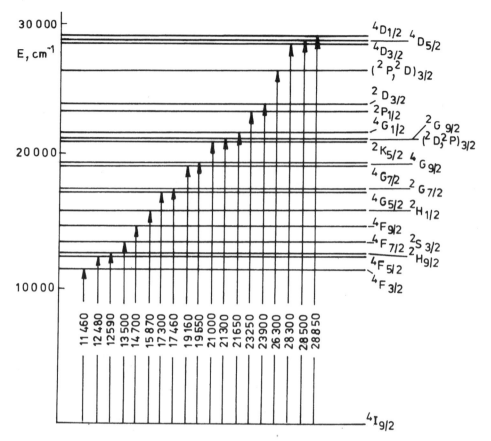

Figure 4.2   Energy levels and electronic transitions in the Nd^{3+} aq ion in the near infrared, visible, and near ultraviolet regions [2].

characteristic. Since at the same time the molar absorption coefficients are often less than $10\ M^{-1}\cdot cm^{-1}$, the oscillator strengths of the individual transitions are about two orders of magnitude less than in the case of d-electronic compounds, at approximately 10^{-5}–10^{-6}.

Thus, the absorption spectrum of a lanthanide compound in solution is characterised by numerous transitions (and correspondingly narrow bands) which are fairly clearly separated from each other. Some of the bands (as well as the areas of maximum light transmittance) are in the 400–700 nm range, which is decisive for the occurrence of a specific colour. As a result of the strong screening of the f electrons (by the $6s$ and $5d$ electrons), the bands are rather insensitive to changes of ligands and the electrostatic field they produce. Consequently, ligands have little effect on the position of bands in comparison with d-electronic elements, also in the visible range. Therefore the colours which occur can be expected to be characteristic for each ion. Table 4.3 gives the colours of Ln^{3+} ions.

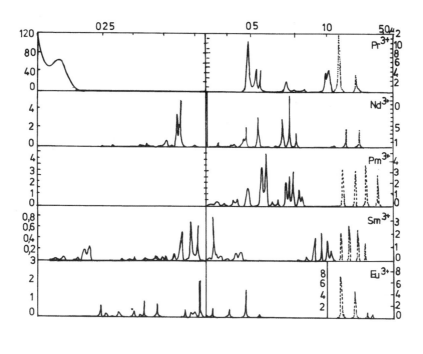

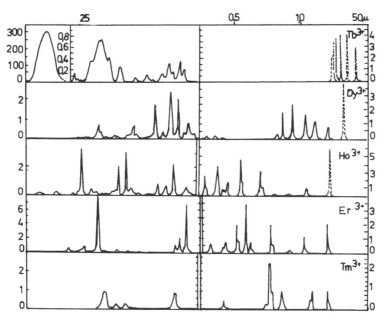

Figure 4.3   Absorption spectra of ten lanthanide ions Ln^{3+} [3].

However, because of different factors, structural or otherwise, the colours corresponding to specific lanthanide ions may be affected. Table 4.4 gives examples

THE COLOUR OF METAL COMPOUNDS

Table 4.3 Colour of Ln^{3+} ions in aqua-complexes

Lanthanide	Number of unpaired electrons	Ground term	Colour
Ce	1	$^2F_{5/2}$	Colourless
Pr	2	$^3H_4$	Green
Nd	3	$^4I_{9/2}$	Lilac
Pm	4	$^5I_4$	Pink
Sm	5	$^6H_{5/2}$	Yellow
Eu	6	$^7F_0$	Very pale pink
Gd	7	$^8S_{7/2}$	Colourless
Tb	6	$^7F_6$	Very pale pink
Dy	5	$^6H_{15/2}$	Yellow
Ho	4	$^5I_8$	Yellow
Er	3	$^4I_{15/2}$	Rose pink
Tm	2	$^3H_6$	Pale green
Yb	1	$^2F_{7/2}$	Colourless
Lu	0	$^1S_0$	Colourless

Table 4.4 Colour of LnX_3 lanthanide halides (in solid state) [4]

	Ce	Pr	Nd	Sm	Eu	Gd	Tb	Dy	Ho	Er	Tm	Yb	Lu
LnF_3	w	g	v	w	w	w	w	g	p	p	w	w	w
$LnCl_3$	w	g	m	y	y	w	w	w	y	v	y	w	w
$LnBr_3$	w	g	v	y	gr	w	w	w	y	v	w	w	w
LnI_3	y	–	g	o	–	y	–	g	y	v	y	w	b

b – brown, g – green, gr – grey, m – mauve, o – orange, p – pink, v – violet, w – white, y – yellow

Table 4.5 Colour of some Ln^{3+} complexes

	tris-(2-propanolate)	tris-(trifluoro-acetylacetonate)	tris-(hexafluoro-acetylacetonate)
Ce^{3+}		dark brown	yellow
Pr^{3+}	green	pale green	pale green
Nd^{3+}	blue	lavender	lavender
Sm^{3+}	pale yellow	cream white	cream white
Eu^{3+}	orange	white	pale yellow
Gd^{3+}	white	white	white
Tb^{3+}	white	white[a]	white[a]
Dy^{3+}	pale yellow	white	cream white
Ho^{3+}	peach	cream white	cream white
Er^{3+}	pink	pink	pink
Tm^{3+}	pale green	white	white
Yb^{3+}	white	white	white
Lu^{3+}	white	white	white

[a] Yellow fluorescence.

of the colour of solid lanthanide halides with the lanthanide in the +3 oxidation state [4]; Table 4.5 gives the colours of some complexes.

As can be seen from Table 4.4, there are some clear differences between different compounds of the same lanthanide. These colour changes to a certain extent resemble the shifts in transition metal halides (Table 3.10) in that the colour shifts bathochromically from the fluoride to the iodide, as for instance in the case of Sm, which may be due to charge-transfer transitions. In other cases, however, e.g. with Dy, the fluoride is green, the chloride and the bromide are white, and the iodide is green again. Table 4.5 shows the differences of colour depending on the ligand (except for Er).

Without going into details, we can say that, as for transition metals, the absorption spectra in the visible range may be due to other electronic transitions or some special features of $f\text{-}f$ transitions. As for $f\text{-}f$ transitions, some of them show increased sensitivity to changes in their environment (ligands) and for this reason are referred to as 'supersensitive' or 'hypersensitive'. They are the subject of many theoretical and experimental studies (e.g. [5, 6, 7, 8]).

Table 4.6 lists the supersensitive transitions occurring in some lanthanides [6]. Except for two transitions in Sm and Dy, all others occur in the visible range, and in the case of Pr, Ho, and Er, there are even two such transitions.

Table 4.6 Supersensitive transitions in electronic spectra of lanthanides [6]

Ln^{3+}	Transition	Wavenumber (cm^{-3})	Wavelength (nm)
Pr	$^3H_4 \rightarrow {}^3P_2$	22500	444
	$^3H_4 \rightarrow {}^1D_2$	17000	588
Nd	$^4I_{9/2} \rightarrow \left.\begin{array}{c}{}^4G_{7/2}\\{}^3G_{19/2}\end{array}\right\}$	19200	521
	$^4I_{9/2} \rightarrow \left.\begin{array}{c}{}^4G_{5/2}\\{}^2G_{7/2}\end{array}\right\}$	17300	578
Sm	$^6H_{5/2} \rightarrow \left.\begin{array}{c}{}^6P_{7/2}\\{}^4D_{1/2}\\{}^4F_{9/2}\end{array}\right\}$	26600	376
	$^6H_{5/2} \rightarrow {}^6F_{1/2}$	6200	1613
Eu	$^7F_0 \rightarrow {}^5D_2$	21500	465
Dy	$^6H_{15/2} \rightarrow {}^6F_{11/2}$	7700	1298
	$^6H_{15/2} \rightarrow \left.\begin{array}{c}{}^4G_{11/2}\\{}^4I_{15/2}\end{array}\right\}$	23400	427
Ho	$^5I_8 \rightarrow {}^3H_6$	28000	357
	$^5I_8 \rightarrow {}^5G_6$	22200	450
Er	$^4I_{15/2} \rightarrow {}^4G_{11/2}$	26500	377
	$^4I_{15/2} \rightarrow {}^2H_{11/2}$	19200	521
Tm	$^3H_6 \rightarrow {}^3H_4$	12600	793

Figure 4.4 shows the absorption spectrum of the Ho^{3+} ion in water-acetone solution and of an $Ho[TTA]_x^{(3-x)+}$ complex ion (TTA is the thenoyl-trifluoroacetylacetonate anion) [6]. The band at 447.5 nm is a supersensitive band, showing an approximately five-fold increase in the molar absorption coefficient, from about 2.5 to about 12.5 $M^{-1} \cdot cm^{-1}$, while other bands in the 400–700 nm range do not undergo changes, whether as to position or as to intensity.

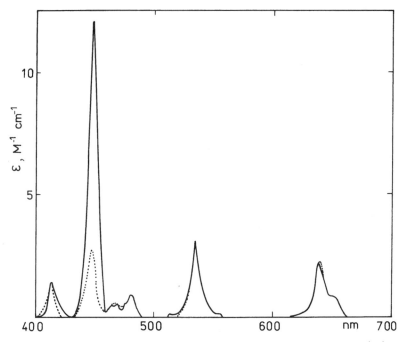

Figure 4.4 Absorption spectra of the Ho^{3+} ion and of its complex $[Ho(TTA)_x]^{(3-x)+}$, TTA = thenoyltrifluoroacetylacetonate, in aqueous acetone (supersensitive transition at 450 nm) [6].

One of the many theories developed to explain the phenomenon of supersensitivity assumes that such transitions are due to the alkalinity of ligands. It has been shown that the oscillator strength of these transitions may be directly correlated with the pK_a value of the ligand, that is the electron donor capacity of the atom bonded directly to the lanthanide. It is important in this connection that this approach can be applied not only to lanthanides in solutions but also to the solid phase, molten salts, and the gaseous phase [6].

Since the question of colour, particularly with respect to the evaluation of colorimetric data, has not been studied for the lanthanides, it is difficult to draw conclusions as to the specific effect of the increase of the intensity of one or even two bands on the colour. In Chapter 3, we discussed the effect of this parameter in the case of a spectrum with one absorption band (in computer simulation). It was found that the colour was much more affected by an increase of the band half-width than by the absorbance value.

So far, only transitions within the f-electronic configuration (in the lanthanides) have been described. In complex compounds with these elements, and particularly with some ligands, charge-transfer transitions have to be considered as well. However, since a charge-transfer (CT) transition is in fact an oxidation-reduction process, it can be anticipated that on the one hand in the case of Ln^{3+} with an unfilled f-subshell an $L \rightarrow M$ transition will be possible, on the other hand in appropriate cases the opposite, $M \rightarrow L$, transitions will occur.

Table 4.7 CT bands in absorption spectra of some $[LnI_6]^{3-}$ complexes [11]

Complex	Wavelength (nm)	Wavenumber ($10^3 \times cm^{-1}$)
$[(C_6H_5)_3PH]_3SmI_6$	402	24.4
	308	32.8
	675	14.8
$[(C_6H_5)_3PH]_3EuI_6$	(450)	(22.8)
	375	26.7
$[(C_6H_5)_3PH]_3TmI_6$	356	28.0
	560	17.85
$[(C_6H_5)_3PH]_3YbI_6$	(445)	(22.4)
	370	27.0

Table 4.8 The first CT band in octahedral halogenide complexes of some lanthanides [9]

	Wavenumbers (cm^{-1})			
	Sm(III)	Eu(III)	Tm(III)	Yb(III)
$[LnCl_6]^{3-}$	43100	33200		36700
$[LnBr_6]^{3-}$	35000	24500	38600	29200
$[LnI_6]^{3-}$	24900	14800	28000	17850

Some data concerning CT transitions in octahedral complexes are given in Tables 4.7 and 4.8 [9, 10, 11]. In accordance with the respective optical electronegativities, in hexahalide complexes we can expect bathochromic shifts of CT bands from Cl to Br to I. The data in Table 4.8 fully support these predictions. Since in the lanthanide complexes under consideration at least one band occurs within the visible range, these CT transitions can have a significant or even decisive effect on the colour observed, as these bands usually are of greater intensity.

Interesting data have recently been obtained in studies of CT transitions in $CaGa_2S_4$ crystals doped with Sm, Tm, Dy, Ho, Nd, and Er (all ions tripositive) [12]. Without going into details, it should be noted that on the basis of Jørgensen's model [13], a CT transition from the sulphur atom (in the matrix) to the $4f$ subshell was detected. For the above six lanthanides, bands with the following wavenumbers were found (in cm^{-1}): 24 500, 26 500, 31 050, 31 100, 32 150, 32 500. Thus, the bands occur in the visible range or on the borderline with the ultraviolet range and affect the colour of these systems.

4.3 SPECTROSCOPIC PROPERTIES AND COLOUR OF LANTHANIDES IN OXIDATION STATES 2+ AND 4+

Although the principal oxidation state for the lanthanides is +3, there are compounds, both in the solid phase and in solutions, where the lanthanides are in the oxidation states mentioned in the heading of this section. Figure 4.5 shows a set of absorption spectra of Ln(II) crystals, from La with an f^1 configuration to Tm (f^{13}), as well as data concerning solutions [14]. As can be seen, practically all ions show

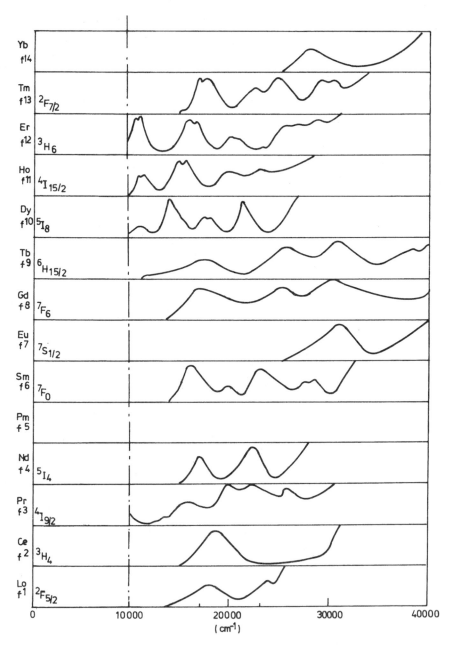

Figure 4.5   Absorption spectra of lanthanide ions Ln^{2+} in CaF_2 matrices (at ambient temperature) [14].

bands in the visible range, and they are shifted most hypsochromically in the case of Eu(II), i.e. for the f^7 electronic configuration. These solutions are either colourless or light yellow green. Sm(II) solutions are blood red, and their spectra do not show absorption below 15 000 cm^{-1}, that is about 660 nm. Also in the spectra of Gd(II) and Tb(II), there are no absorption bands below that value, although they differ from each other in the shorter wave range.

The main characteristic to be noticed in the spectra under discussion is the greater half-width of the absorption bands. On this basis, transitions in Ln^{2+} ions are interpreted as f-d transitions, especially in the lower energy range. Fluorescent transitions are interpreted as f-f transitions, just as in the case of tripositive lanthanides.

As for the +4 oxidation state, Ce(IV) compounds have been studied most thoroughly. As in this case it is a closed-shell configuration (f^0), electronic transitions can be caused by CT transitions. Most Ce(IV) compounds are yellow or orange, and their absorption spectra show wide and intensive bands in near ultraviolet. In the $CeCl_6^{2-}$ ion, a more intense band has been found at $26\,600\,cm^{-1}$ (376 nm, [4]). It has also been found that a 0.01 M solution of crystalline $(NH_4)_2[Ce(NO_3)_6]$ in $HClO_4$ (1–3 M) is initially yellow and after several hours becomes orange. An intense purple colour of an ion also supports the CT nature of a transition.

Some coloured Nd(IV) systems are known, with orange Cs_3NdF_7 being the most fully studied. The first CT band occurs at approximately 385 nm, but the colour is also affected by f-f transitions (in the $4f^2$ configuration), which occur in the visible range [15]. The colours of Ln(II) halides are given in Table 4.9 [4].

Table 4.9 Colour of LnX_2 lanthanide halides (in the solid state) [4]

Ce	CeI_2	bronze
Pr	PrI_2	bronze
Nd	$NdCl_2$	green
	NdI_2	violet
Sm	SmF_2	purple
	$SmCl_2$	brown
	$SmBr_2$	brown
	SmI_2	green
Eu	EuF_2	greenish yellow
	$EuCl_2$	white
	$EuBr_2$	white
	EuI_2	green
Gd	GdI_2	bronze
Dy	$DyCl_2$	black
	DyI_2	purple
Tm	$TmCl_2$	green
	TmI_2	black
Yb	YbF_2	grey
	$YbCl_2$	green
	$YbBr_2$	yellow
	YbI_2	black

4.4   USE OF LANTHANIDES AS COLOURED SUBSTANCES

As discussed above, the colour of the lanthanides, especially ions and compounds in the +3 oxidation state, is mostly due to f-f transitions and narrow absorption bands, whose position and intensity do not change much (except for 'supersensitive' transitions) with change of ligands in complex and simple compounds. This principal feature is the basis for the use of lanthanides as coloured substances.

From the point of view of colorimetry, there are two possible directions of application, namely in the phenomena of additive and subtractive mixing of colours or colour substances. We shall describe some aspects of such applications on the basis of, amongst others, Maestro's review [16].

Obtaining a specific colour through subtractive processes is used mostly in glass colouring with the use of rare-earth oxides dissolved in glass. In comparison with transition metal compounds, such as titanium, vanadium, and chromium compounds, rare-earth oxides show both narrow absorption bands and good thermal resistance under the technological conditions of glass production. This facilitates obtaining a permanent colour. One of the most frequently used compounds is Nd_2O_3. The colour of glass doped with that oxide may change from purple to blue depending on the concentration of the oxide and also due to dichroism. As a result of strong light absorption in the yellow and orange ranges of the spectrum, glass containing neodymium oxide is used as a filter, and now also to enhance the contrast of colour television screens.

Apart from neodymium oxide, praseodymium and erbium oxides are also used. As can be expected, the presence of praseodymium oxide causes a green colouring of the glass, while erbium oxide brings about a pink colour, associated with a band at approximately 530 nm. However, the high price of the latter substance limits its use.

The permanence of colours obtained with lanthanides is also used in the technology of coloured ceramic products. One of such applications is the use of praseodymium oxide in a zirconium matrix: extremely bright yellow pigments are obtained.

The use of rare earths in additive colour mixing processes is to do with their luminescence. One of the most important fields of application is colour television, where yttrium-europium red is used. Another field is radioluminescence, where – depending on the spectral sensitivity of the emulsion – blue or green emitters are required. The $BaFCl:Eu^{2+}$ system is a good emitter in the blue and ultraviolet range, while $Gd_2O_2S:Tb^{3+}$ and LaOBr:Tb or Tm^{3+} are used in the case of blue and green emulsions.

The review cited above [16] stresses that the permanence and purity of the colours produced permits the use of compounds of the rare earths to transform ultraviolet radiation into visible light through the use of luminophores. One of the substances used for this purpose is $YVO_4:Eu^{3+}$, which compensates the lack of red light emission in high-pressure mercury lamps. Lanthanide luminophores are also used to transform the ultraviolet radiation of low-pressure mercury lamps into visible light with the use of blue green and red emitters. An important purpose is to obtain high luminescence efficiency and a high colour-rendering index. Three entities have been used: $BaMgAl_{10}O_{20}:Eu^{2+}$ with blue emission at 450 nm, $LnMgAl_{11}O_{19}:Tb^{3+}$ with Ce^{3+} as an activator, which gives green emission at 545 nm, and $Y_2O_3:Eu^{3+}$ with red emission at 611 nm.

It should be noted that the measurement of chromaticity coordinates, e.g. by the CIE method, is used to determine the optimum composition of the emitter, essential to obtain the appropriate colour and purity [16].

REFERENCES

[1]   Carnall, W.T., Fields, P.R. and Rajnak, K., 1968, *J. Chem. Phys.*, **49**, 4424.
[2]   Crosswhite, H.H., Dieke, G.H. and Carter, W.J., 1965, *J. Chem. Phys.*, **43**, 2047.
[3]   Carnall, W.T., Crosswhite, H. and Rajnak, K., 1985, in *Rare Earth Spectroscopy*, ed. B. Jeżowska-Trzebiatowska, J. Legendziewicz, and W. Stręk, World Scientific, Singapore, p. 267.
[4]   Greenwood, N.N. and Earnshaw, A., 1998, *Chemistry of the Elements*, 2nd edn., Pergamon, Oxford.
[5]   Carnall, W.T., Fields, P.R. and Wybourne, B.G., 1965, *J. Chem. Phys.*, **42**, 3797.
[6]   Henrie, D., Fellows, R.L. and Choppin, G.R., 1976, *Coord. Chem. Rev.*, **18**, 199.
[7]   Judd, B.R., 1962, *Phys. Rev.*, **127**, 750.
[8]   Ofelt, G.S., 1962, *J. Chem. Phys.*, **37**, 511.
[9]   Jørgensen, C.K., 1988, in *Handbook of the Physics and Chemistry of Rare Elements*, Vol. 11, Chapter 75, North-Holland, Amsterdam, p. 197.
[10] Ryan, J.L. and Jørgensen, C.K., 1969, *J. Phys. Chem.*, **70**, 2845.
[11] Ryan, J.L., 1969, *Inorg. Chem.*, **8**, 2053.
[12] Garcia, A., Ibanez, R. and Fouassier, C., 1985, in *Rare Earth Spectroscopy*, ed. B. Jeżowska-Trzebiatowska, J. Legendziewicz, and W. Stręk, World Scientific, Singapore, p. 412.
[13] Jørgensen, C.K., 1962, *Mol. Phys.*, **5**, 271.
[14] Conway, J.G., 1985, in *Rare Earth Spectroscopy*, ed. Jeżowska-Trzebiatowska, B., Legendziewicz, J. and Stręk, W., World Scientific, Singapore, p. 3.
[15] Varga, L.P. and Asprey, L.B., 1968, *J. Chem. Phys.*, **48**, 139.
[16] Maestro, P., 1985, *J. Less-Common Met.*, **111**, 43.

5. COLOUR IN CHEMICAL ANALYSIS

The colour of a chemical substance is one of its main characteristics, along with its general appearance, the form in which it occurs, and its smell, which are directly observable. If there is no information available as to whether the substance is organic, inorganic, an inorganic complex with organic ligands, or an organometallic compound, then colour information is limited simply to predictions of ranges in the visible region in which light absorption may occur. Because of the phenomenon of metamerism (Chapter 1), even these predictions can be imprecise, or even misleading.

Metal compounds provide the main focus for this book, with a special emphasis on complexes of the d- and f-block elements. If we also consider colours of compounds of s- and p-block elements, then the overall situation may be summarised as shown in Table 5.1.

Thus if metal compounds are the object of study, colour may serve as preliminary information. In order to simplify our exposition, we shall restrict our attention to the colour of compounds (ions) of d- and f-electronic elements. The simplest situation occurs if the analysed substance is known *a priori* to contain one or more chemical

Table 5.1 Colours of compounds and complexes around the Periodic Table

	s	p	d	f	
				$4f$	$5f$
Colour of ion/salts[a]	colourless	colourless	coloured	weakly coloured	weakly coloured
Colour of complexes[a]					
with inorganic ligands			strongly ligand-dependent	usually weakly ligand-dependent	partly ligand-dependent
with organic ligands	←——————— frequently strongly coloured ——————→				
	(often with high or very high molar absorption coefficient)				

[a] Colours of salts and complexes will of course be greatly affected if the counterion or ligand is itself strongly coloured.

131

compounds of these groups of elements. In such a case the perceived colour may be a certain source of information, which, however, without further chemical and physicochemical investigation, is only approximate.

In Chapter 3, we discussed these issues in more detail, using – among other things – the example of the computer simulation of the colour of chromium compounds, and we showed that there is a clear relationship between colour and ligand field strength.

One of the aspects which are well known in practice is the subjective description of colour with an appropriate name. This is particularly important in describing new compounds in the solid phase. If we disregard such factors as lighting (time of the day) and environment, we find that colour can change depending on grain size, which is very rarely mentioned in publications. The issue becomes even more complicated if we compare terminology used in different languages.

Below we shall discuss the use of colour in qualitative analysis, quantitative spectrophotometric analysis, structural chemical analysis, and the use of quantitative colour measurements (colorimetry) in some analytical applications.

5.1   COLOUR IN QUALITATIVE ANALYSIS OF METAL COMPOUNDS AND IONS

This topic is familiar to every chemist, as it is an important stage in getting to know the properties and reactions of each metal ion and its derivatives. Solubility and colour are the two key properties which underlie the separation of metals into groups, and their subsequent specific identification, in the classical scheme of qualitative inorganic analysis. The subject of solubility is discussed in depth in many analytical textbooks, but an in-depth treatment of the colours involved is very rarely encountered.

Some of the group reagents produce clear and characteristic colour reactions or coloured precipitates with metal cations (or other groups). One such group reagent is $(NH_4)_2S$ in the presence of $NH_4OH + NH_4Cl$, which forms coloured sulphides (or hydroxides) with the cations in group III of the standard scheme of qualitative analysis. In addition to group reagents, an important role is also played by the element-specific reagents which give characteristic coloured precipitates or solutions for individual cations.

Table 5.2 [1] presents a selection of the main reagents used in identifying transition metal cations. Except in a few cases, coloured precipitates or solutions are obtained, which are characteristic of the respective metal ions. This Table also provides a good illustration of the occurrence of different colours in compounds consisting of a given cation in range of different ligand environments.

Naturally, ideally we would like to be able to use selective reagents, possibly reacting with only a single cation (so-called specific reagents), or with just a few but then only under strictly defined analytical conditions.

Characteristic colour reactions are also known for s- and p-electronic cations. In that case, however, the reagents have their own colour or special properties. Thus, for instance, the K^+ ion reacts with the brown $Na_3[Co(NO_2)_6]$ solution, producing

Table 5.2 Some analytical reactions of Group III cations

	Al^{3+}	Cr^{3+}	Fe^{3+}	Fe^{2+}	Mn^{2+}	Zn^{2+}	Co^{2+}	Ni^{2+}
$(NH_4)_2S$	White pptt[a]	Dull green pptt[bc]	Black pptt[c]	Black pptt[c]	Flesh coloured pptt[cd]	White pptt[c]	Black pptt[c]	Black pptt[c]
KOH; NaOH	White pptt[a]	Dull green pptt[b]	Dark brown pptt	White pptt[e]	Brownish white pptt	White pptt	Blue pptt	Green pptt
KOH; NaOH plus H_2O_2		Bright yellow colour[f]	Dark brown pptt	Dark brown pptt[g]	Dark brown pptt	White pptt	Black pptt[g]	Green pptt
$K_4[Fe(CN)_6]$			Dark blue pptt	White pptt, turning blue	White pptt	White pptt	Green pptt	Pale green[h] pptt
$K_3[Fe(CN)_6]$				Dark blue pptt	Brown pptt	Light yellow pptt	Deep red pptt	Yellowish pptt
Borate/phosphate bead test		Green	Brown		Violet		Blue	Brown

[a] $Al(OH)_3$; [b] $Cr(OH)_3$; [c] sulphide; [d] slowly turns dark green (see N.V. Sidgwick, *The Chemical Elements and Their Compounds, Vol. 2*, Clarendon Press, Oxford, 1950, p. 1284 for the chemistry involved); [e] turning brown, except in the absence of air; [f] CrO_4^{2-}; [g] hydrated M_2O_3; [h] eau de nil.

a yellow precipitate of potassium hexanitrocobaltate(III), while the Ba^{2+} ion in reaction with coloured $K_2Cr_2O_7$ forms a yellow precipitate of $BaCrO_4$.

Selective and specific reagents play an important role in the qualitative and quantitative analysis of the lanthanides. As their own coloration in the +3 oxidation state is weak, i.e. their molar absorption coefficients are small (cf. Chapter 4), the reagents used for these elements form strongly coloured compounds (e.g. dyes).

Qualitative identification of chemical substances by means of colour reactions (or precipitates) has been used since the early days of analytical chemistry. Such reactions and properties have also played an important role in materials science, practically in all fields, such as decorative art, jewellery making, painting, architecture, food evaluation, mineral and gemstone identification and many others. Gradually the colour of various objects has become the subject of much deeper investigations, whose main purpose has been the interpretation of spectra in the visible range and the use of colour for the purposes of qualitative analysis.

5.2   SPECTROPHOTOMETRIC QUANTITATIVE ANALYSIS

The spectrophotometric method of element determination, including determination of metals, consists of using coloured compounds (complexes), which can be formed by the examined elements with appropriate reagents (ligands). The basis of such analysis is a definite dependence of the absorbance of the solution on the concentration of the coloured substance. In a favourable case, according to the Bouguer-Lambert-Beer law, the dependence is linear. It may however be a curve, with a deviation towards higher or lower ordinate values (absorbance). Deviations from a straight line may occur, for instance, for particular concentrations or for a particular range of pH values. They also depend on physical factors, for example if a particular radiation is not monochromatic. Without going into details, which are described in the basic spectrophotometric analysis textbooks, it should be noted that the preparation (or knowledge) of a suitable calibration curve enables the use of the described method for the purpose of quantitative analysis even in such cases.

Spectrophotometric methods are precise and sensitive [2] – indeed afford the most precise methods of instrumental analysis. The literature on this subject is very rich; many references and descriptions of spectrophotometric analyses may be found in [3]. On the basis of this source and primary research papers, Table 5.3 gives a selection of colour reactions used in the spectrophotometric determination of $3d$ transition elements. To complement this Table, Table 5.4 shows how colours vary with the nature of the metal being determined in the specific case of dithizone analyses. The spectra of this ligand and three of its complexes are shown in Figure 5.1.

The analysis of d-electronic elements (and particularly $3d$-electronic) could make use of aqua-ions which are coloured (or other systems). However, small molar absorption coefficients in practice preclude the use of aqua-ions, especially since complexes with much greater indices can be used. It is also believed that in sensitive spectrophotometric methods the ε_{max} value of the coloured compound should be greater than $1.10^4 \, M^{-1} \cdot cm^{-1}$, and a value less than 1.10^3 characterises rather less sensitive methods. For this reason, inorganic reagents, which usually form weakly

Table 5.3 Some colour reactions for spectrophotometric determination of 3d-elements

Element	Reagent (reaction)	Colour of solution or precipitate	Wavelength (nm)	Molar extinction coefficient	
Ti	chromotropic acid	brown red	460	1.7×10^4	pH 3.5 (colour strongly pH-dependent)
	hydrogen peroxide	yellow orange	410	7×10^2	
V	8-hydroxyquinoline	yellow	550	3.0×10^3	in chloroform
	N-Bz-N-Ph-hydroxylamine[a]	violet	525	5.1×10^3	in chloroform
Cr	1,5-diphenylcarbazide	violet	546	4.2×10^4	
	oxidation to dichromate	orange	350	7.5×10^2	
	oxidation to chromate	yellow	373	1.4×10^3	
	edta	violet	540	1.4×10^2	pH 4 to 5
Mn	oxidation to permanganate	violet	528	2.4×10^3	0.5 M H_2SO_4
	1-(2-pyridylazo)-2-naphthol	violet red	564	5.8×10^4	pH 9.2
Fe	thiocyanate	red	470–530	8.5×10^3	Fe^{3+}; in water
				1.8×10^4	Fe^{3+}; in water/acetone
	1,10-phenanthroline	orange red	512	1.1×10^4	Fe^{2+}
	2,2'-bipyridyl	pink	522	8.7×10^3	Fe^{2+}
Co	1-nitroso-2-naphthol	orange	415	2.9×10^4	oxidation $Co^{2+} \rightarrow Co^{3+}$
	2-nitroso-1-naphthol	brown red	365	3.7×10^4	oxidation $Co^{2+} \rightarrow Co^{3+}$
	thiocyanate	blue → pink	620	1.9×10^3	colour is concentration dependent
Ni	dimethylglyoxime	yellow(sol); red (pptt)	546	3.4×10^3	see footnotes [b] and [c]
	quinoxaline-2,3-dithiol	blue	660	1.8×10^4	in acid solution
	1-(2-pyridylazo)-2-naphthol	violet	565	5.3×10^4	
Cu	cuproin	purple	546	6.4×10^3	$Cu^{2+} \rightarrow Cu^+$; in isoamyl alcohol
	neocuproin	orange	454	7.9×10^3	
	bathocuproin	orange	479	1.4×10^4	

[a] Bz = benzyl, Ph = phenyl. [b] $\lambda_{max}/\varepsilon$ data are given for chloroform solution. [c] The solid and solution colours differ because intermolecular interactions present in the solid do not occur in solution, though when $Ni(dmgH)_2$ is solubilised in micelles or microemulsions it is red – presumably the solubilised form contains fragments of the stacks which characterise the solid.

Table 5.4 Colours of dithizone complexes encountered in analytical chemistry

Metal	λ_{max}	Colour
Selenium	420	yellow
Gold	420	yellow-brown
Silver	462	orange-yellow
Mercury	485	orange-yellow
Bismuth	490	orange-brown
Indium	~510	pink
Cadmium	520	pink
Lead	520	pinkish-red
Zinc	538	pink
Copper	550	violet
Palladium	450, 640	grey-green
Platinum	490, 710	brown-yellow
Ligand	450, 620	green[a]

[a] The shade of green depends somewhat on the solvent.

coloured complexes, are seldom used. Most reagents are organic compounds, often colourless (at least under the determination conditions), but also coloured, for instance dye-type reagents. Many organic reagents form strongly coloured chelate complexes with transition metal ions.

An important issue in quantitative analysis using colour reactions is the permanence of particular colours. This means that during measurement (or between

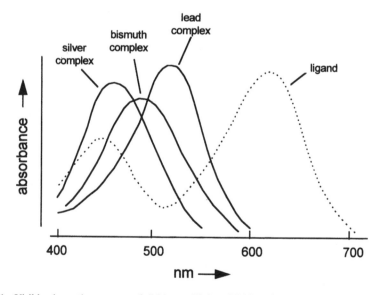

Figure 5.1 Visible absorption spectra of dithizone (diphenyldithiocarbazone, Hdtz) and its silver, lead, and bismuth complexes, Ag(dtz), Pb(dtz)$_2$, and Bi(dtz)$_3$, in carbon tetrachloride solution.

several measurements) the molar absorption coefficient at a given wavelength cannot undergo any changes. Such changes can be caused, for instance, by the gradual formation of successive complexes, dissociation-association reactions, or partial precipitation. Thus each of such colour reactions must be thoroughly studied taking into account various experimental and theoretical aspects.

5.3 ROLE OF COLOUR IN STRUCTURAL–CHEMICAL CONSIDERATIONS

In Chapter 3, as well as in the chapters devoted to the colour of glasses and pigments, we presented both the results of simulation studies and experimental results which show certain structural data can be correlated with the colour of a chemical compound or, more generally, of the system which is being studied.

One of such parameters is the metal-oxygen distance in oxides. Particularly rich experimental material has been gathered in studies of the colour of Cr_2O_3 (and to a certain extent also NiO), which was discussed in Chapter 3. It is interesting that the dependency has also been found to hold in minerals. The usefulness of this parameter is understandable in the light of ligand field theory, where there is a definite dependency between metal-ligand atom distance and the value of the Dq parameter. However an important problem lies in the generally recognised fact, stressed repeatedly in this book, that the colour–structure relationship is basically unidirectional and possible bidirectionality can only be predicted in a certain narrow and limited range of cases.

Computer simulation for Cr(III) has shown a very good correlation between the colour computed for a given concentration of metal ions and the thickness of the light-absorbing layer (i.e. concrete absorbance) on the one hand and the Dq parameter on the other.

The use of the above dependency to predict the colour of a solid-phase substance may be difficult or even deceptive if the crystal structure is disturbed or if there are impurities. However the main reason is the fact that many compounds, particularly transition metal compounds, owe their colour not only to ligand field transitions but also to CT transitions, and sometimes only the latter are the immediate cause of the occurrence of colour. This is of course due to the much greater intensity of CT transitions in comparison with ligand field transitions. Both transition metal and lanthanide halides serve as good examples (Tables 3.12–3.14 and 4.4).

The colour of solutions of d- and of f-electronic metal compounds also depends on some structural aspects. The most significant seems to be the possibility of the occurrence of coordination and configuration equilibria: this issue was discussed in Section 3.1.1 devoted to solvatochromism. Even the effect of metal (ion) concentration in the solution cannot be unequivocally predicted – this has also been discussed (dichromatism).

In the case of the colour of solutions it seems worthwhile and useful to make wider use of colorimetric methods, that is quantitative colour measurements, e.g. CIE, CIELAB or CIELUV.

5.4   INDICATORS

Colour and colour differences, or more generally chromaticity and its changes, are of central importance in the selection of indicators for volumetric titrations with visual end-point determination. Although instrumental methods, such as potentiometric or conductometric titration, are commonly used in chemical analysis, in many cases colour-change indicators are simpler to use, or given more precise results [4, 5]. Examples include acid-base titrations in non-aqueous environments, especially in solvents of low permittivity such as acetic acid. Complexometric titrations provide another important area, arguably more relevant in the context of the present book, where colour-change indicators are very widely used.

The use of tristimulus colorimetry, in accordance with the 1931 CIE colour specifications, in the colour description and selection of indicators has recently been reviewed in detail [6]. The discussion of complexometric titrations included mention of the assessment of Arsenazo III for titration of a selection of metals, and of the assessment of ranges of indicators for such determinations as that of lead(II), of calcium(II), of magnesium(II), of zinc(II), and of bismuth(III) with EDTA. This review also deals with screened indicators, introducing the parameter 'relative greyness' to further refine the assessment and characterisation of the quality of colour changes at end-points. Tristimulus colorimetry has proved particularly useful in the important but complicated problem of the analysis of plutonium-containing solutions, where the four oxidation states of 3+, 4+, 5+, and 6+ can coexist in equilibrium [7].

5.5   CHROMATICITY OF INDICATORS

We shall start our discussion of the chromaticity of indicators by dealing with some organic compounds used in non-aqueous acid-base titrations. Their colour changes may be characterized by means of trichromatic colorimetry – Figures 5.2, 5.3, and 5.4 [4] show results for the three indicators methyl violet, alizarin-9-imine, and quinalizarin-9-imine used in an environment of glacial acetic acid. The first of these three Figures shows the CIE x, y chromaticity diagram with chromaticity traces for all three indicators. As can be seen, methyl violet changes colour from violet to blue, and then from blue to yellow, while the other two indicators show a change from red to yellow. The changes are even more clearly seen in the CIELAB chromaticity diagram (Figure 5.3).

Figure 5.4, in the $\Delta E / \Delta pc_{HClO_4} - pc_{HClO_4}$ coordinate system, enables the determination of the sensitivity of the indicators used (for the optimum indicator concentration). As can be seen, the sensitivity of colour changes is greatest for methyl violet. Let us recall that ΔE is given by the following equation:

$$\Delta E_{ab}^* = \left[(\Delta L^*)^2 + (\Delta a^*)^2 + (\Delta b^*)^2 \right]^{1/2}.$$

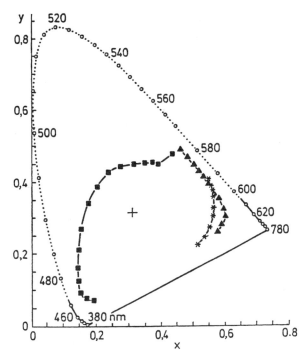

Figure 5.2   CIE x, y chromaticity coordinates of three indicators (at optimal concentrations): ▲ alizarin-9-imine; * quinalizarin-9-imine; ▪ methyl violet [4].

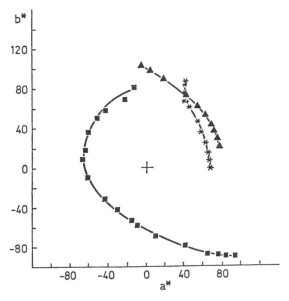

Figure 5.3   CIE a^*, b^* chromaticity coordinates (at optimal concentrations) of the three indicators shown in Figure 5.2 [4].

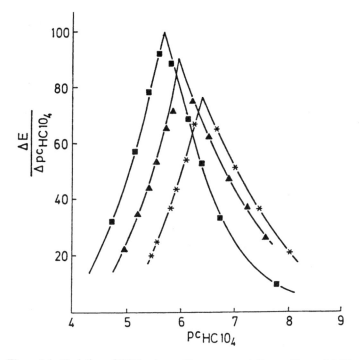

Figure 5.4   Variation of SCD values with pc_{HClO_4}; notation as Figure 5.2 [4].

Many publications have been devoted to chromaticity changes in acid-base systems in non-aqueous solvents. Thus, e.g. dimethylformamide has been used as an aprotic solvent of intermediate permittivity [8]. As is well known, this solvent is widely used in coordination chemistry and in studies of solvatochromism. CIELAB chromaticity coordinates were determined as well as so-called Qy-Qx complementary system chromaticity coordinates [9], pH values for the maximum colour change, and the sensitivity and range of colour change for various indicators.

The literature also provides information about a computer program which computes the above-mentioned values characterising chromaticity in different cases [10]. A diagram of the program is given in Figure 5.5.

The use of indicators in chemical analysis is a classical issue, but it seems that the use of chromaticity and its changes has not become very popular yet.

Another kind of coloured indicators are transition metal complexes, which can be used to evaluate the acid-base properties of solvents. This issue was partially discussed in Chapter 3, where the effect of solvent on metal complex colours was considered. This phenomenon is termed solvatochromism, of which some aspects were considered in Chapter 3.

The use of transition metal complexes as coloured indicators dates back to the 1970s [11]. Although solvent effects on colour have long been known and studied, the use of colour to evaluate the properties of solvents requires a different approach and perspective. Twenty years ago only a dozen or so transition metal complexes were recognised as potential indicators of the donor-acceptor properties (Lewis acidity) of

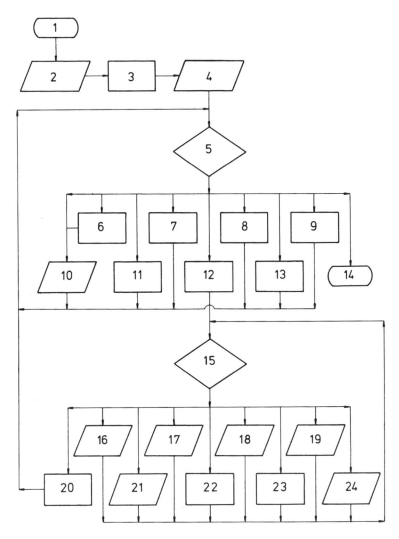

Figure 5.5 Flow chart for the SUPERCOLOR program:

1 Start
2 Choose illuminant and observer
3 State constants
4 Select indicators and symbols
5 Option 1
6 Calculate coordinates
7 Calculate screened indicators
8 Calculate pK
9 Calculate optimum concentration
10 Print results
11 Printer on/off
12 Graphics

13 Change indicators and symbols
14 End
15 Option 2
16 $Q_y$vs.Q_x; y vs. x; v' vs. u'
17 Lab, $U^*V^*W^*$; CIELAB; CIELUV
18 ΔE or g pH or q and derivatives
19 pH/ΔE vs. log (C/C_{ref})
20 Return to option 1
21 Spectra (A or T)
22 Plotter on/off
23 Clear graphics screen
24 Print graphics screen

Table 5.5 Colours of [Fe(phen)$_2$(CN)$_2$] solutions in solvents covering a range of acceptor numbers, AN

Solvent	AN	Colour
hexamethylphosphortriamide	11	blue
dimethylformamide	16	violet blue
dichloromethane	20	violet
ethanol	37	ruby
glacial acetic acid	53	light red
formic acid	84	yellow
trichloroacetic acid	105	light yellow

solvents [12], but in recent years many more such complexes have been synthesised and examined in this connection (cf. Chapter 3). An essential feature of such transition metal complexes and organometallic compounds is their incorporation of ligands with a lone pair of electrons capable of electrophilic solvation. Such ligands include, for instance, carbonyl and cyanide. Examples of complexes include [Mo(bipy)(CO)$_4$] and its tungsten analogue, and [Fe(bipy)$_2$(CN)$_2$], [Fe(bipy)(CN)$_4$]$^{2-}$ and their phen analogues. The last-named has particularly suitable properties and is frequently used.

It can be inferred that, in accordance with the Lewis theory of acids and bases, some metal complexes may act as basic indicators with respect to non-aqueous solvents having a substantial capacity to act as electron pair donors. Sone and Fukuda [13] demonstrated this for the square-planar copper(II) complex [Cu(tmen)(acac)]$^+$, where tmen = N,N,N′,N′-tetramethylethylenediamine and acac = acetylacetonate. The colour of this complex depends very clearly on the donor properties of the respective solvents; the wavenumbers of the absorption band maximum are linearly dependent on Gutmann donor numbers, DN. Colour differences are very marked, and indeed it is possible to estimate DN values visually [12].

Moving from colour to electronic transitions, the intense colour of the acidic indicator [Fe(phen)$_2$(CN)$_2$] is assigned to a $t_{2g} \rightarrow \pi^*$ metal-to-ligand charge-transfer (MLCT), whereas the considerably less intense colour of its iron(III) analogue [Fe(phen)$_2$(CN)$_2$]$^+$ may be attributed to a d-d transition, perhaps with some contribution from relatively unfavourable ligand-to-metal charge-transfer (LMCT). The colour of the basic indicator [Cu(tmen)(acac)]$^+$ may also be assigned to a ligand field (d-d) transition.

We shall now see how the perceived colour of investigated indicators changes with the nature of the solvent. The absorption spectra of both types of iron complexes acting as acidic indicators have been described [12]. The colours of solutions of [Fe(phen)$_2$(CN)$_2$] in a range of solvents are given in Table 5.5, together with the acceptor numbers, AN, of the solvents.[1] Without going too far into details – this type

[1] The full range of colours exhibited by Fe(phen)$_2$(CN)$_2$, from yellow in acid aqueous solution to blue in solvents of low acceptor numbers (AN), can be seen in the relevant colour plate in H.W. Roesky and K. Möckel, *Chemical Curiosities*, VCH, Weinheim, 1996 (see Plate 16); solvent effects on the colour of solutions of an analogous Schiff base complex are shown in Figure C3.

of complex was partially discussed in Chapter 3 – it should be noted once again that the colours indicated in Table 5.4 are defined only qualitatively and, what is more important, the position of only one band is taken into consideration. The intensities of bands and the overall contour are disregarded (Figure 5.6) [14]. In recent studies

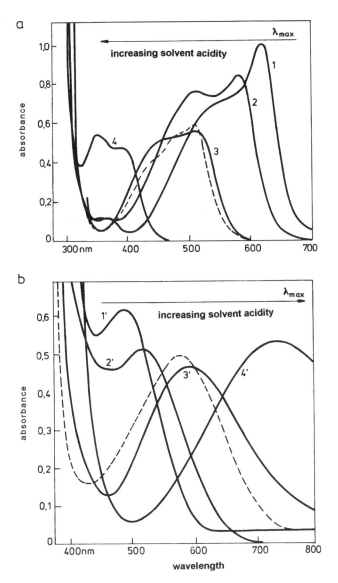

Figure 5.6 Solvent-dependent spectra of $Fe(phen)_2(CN)_2$ (a) and of $Fe(phen)_2(CN)_2^+$ (b). (a): $[Fe(phen)_2(CN)_2] = 1 \times 10^{-4}\,mol\,dm^{-3}$, in: 1 – dimethylformamide, 2 – nitromethane, 3 – dilute hydrochloric acid, and 4 – concentrated sulphuric acid; broken line $[Fe(phen)_3^{2+}] = 5 \times 10^{-5}\,mol\,dm^{-3}$, in acetonitrile. (b): $[Fe(phen)_2(CN)_2^+] = 5 \times 10^{-4}\,mol\,dm^{-3}$, in: 1' – nitromethane, 2' – formic acid, 3' – 70% perchloric acid, and 4' – concentrated sulphuric acid; broken line $[Fe(phen)_3^{3+}] = 5 \times 10^{-4}\,mol\,dm^{-3}$ in acetonitrile. Path length = 1 cm in all cases.

[15] it was shown that h_{ab} does clearly depend on solvent acidity parameters, expressed as AN, Z, or E_T (Figure 5.7).

In regard to solvatochromism, we stated that in principle various physicochemical methods may be used to determine the effect of, and possible correlations with, solvent properties. Thus, the determination of colour and of colour shifts is only one of the methods used for this purpose, where the colour is defined qualitatively and at the same time the magnitude of the changes of various parameters of electronic spectra in the range 380 to 780 nm are evaluated.

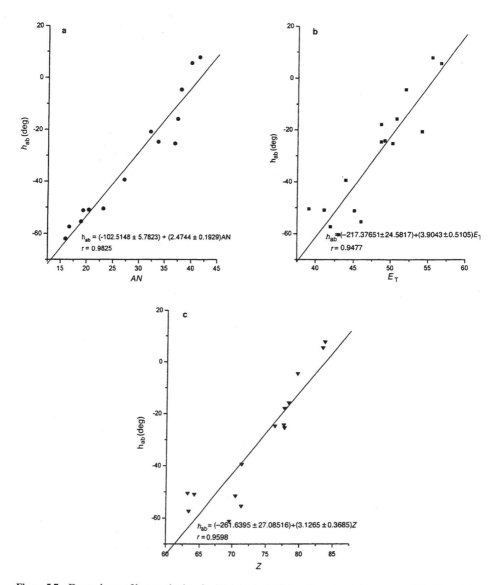

Figure 5.7   Dependence of hue angle, h_{ab}, for [Fe(phen)$_2$(CN)$_2$] on the solvent parameters AN, E_T, and Z.

It should be noted, however, that the use of these two sources does not correspond to the same situation. As is well known, colour is due to the overall contour of the spectrum including both the positions and intensity distribution of absorption bands. But when the acid properties of solvents, characterised for instance by their acceptor numbers, are compared with the colour of a complex in the respective solvents, only one number is used, viz. the wavenumber of a single absorption band. However, the energy of this one transition is not the only parameter determining the colour of the solution – as demonstrated in Chapter 3 and shown in Figure C2, it is possible for two solutions to have identical ν_{max} values but significantly different colours. Further, the wavenumber taken directly from a spectrum print-out or plot does not correspond exactly with the actual transition energy because of absorption band interference and the occurrence of shoulders. If the intensity of the band corresponding to a given transition changes from solvent to solvent, then colour shifts cannot be directly correlated with the energy of this electronic transition.

When discussing the possibility of predicting colours of solutions of transition metal complexes (see Chapter 3), we concluded that if the envelope and the intensity of absorption spectra are preserved, the fundamental condition is only the change in the position of the band or bands. In such cases Dq is an essential parameter, as has proved to be correct in the case of chromium(III) and nickel(II) complexes.

Bartecki and Tłaczała [16, 17] conducted chromaticity studies for several complexes in water and in organic solvents. The values of CIE and CIELAB chromaticity coordinates for $CoCl_2$ and for $CoBr_2$ in some solvents are given in Tables 5.6 and 5.7; Table 5.8 gives values of ΔE_{ab}^* and ΔH_{ab}^* for solutions of $CoBr_2$ in a series of organic solvents relative to aqueous solution. These studies represent a further step towards a more precise evaluation of colour shifts of complexes, and an attempt to correlate such shifts with processes occurring in the complex-solvent system.

Cobalt(II) complexes may act as indicators of changes in coordination geometry, as the two main forms, octahedral and tetrahedral, may both occur in solutions. This matter has been discussed for many years, but quantitative measurements of colour have not yet been used for this purpose.

Table 5.6 CIE and CIELAB chromaticity coordinates of $CoCl_2$ in organic solvents[a] ($c = 0.01$ M, $d = 1$ cm) [16]

Solvent	x	y	Y	$\lambda_d(\lambda_c^*)$ (nm)	L*	a*	b*	h_{ab} (deg)
MeOH	0.3199	0.3216	90.3806	518*	112.154	7.374	−1.5623	348.33
F	0.3278	0.3144	83.7994	507*	109.353	14.770	−2.804	349.33
NMF	0.3282	0.3153	81.0732	502*	108.164	14.350	−2.340	350.74
DMSO	0.1933	0.2871	51.0462	487	92.707	−43.386	−29.479	214.19
DMF	0.1796	0.2729	44.1424	486	88.324	−43.324	−34.895	218.07
AN	0.1518	0.2424	36.0045	485	82.523	−46.194	−45.421	224.52
A	0.1445	0.2126	22.5273	482	70.582	−32.179	−49.376	236.91
i-PrOH	0.1405	0.1802	22.7138	480	70.776	−19.450	−62.592	252.74
t-BuOH	0.1452	0.1790	23.3473	479	71.428	−15.811	−63.254	255.95

[a] Solvent abbreviations: F = formamide; NMF = N-methylformamide; AN = acetonitrile; A = acetone.

THE COLOUR OF METAL COMPOUNDS

Table 5.7 CIE and CIELAB chromaticity coordinates of 0.01 M CoBr$_2$ solutions in various solvents ($d = 1$ cm)

Solvent[a]	x	y	Y	L^*	a^*	b^*
H$_2$O	0.3197	0.3286	93.8373	113.566	3.819	1.115
MeOH	0.3196	0.3216	90.7104	112.291	7.236	−1.557
Me$_2$CO	0.1644	0.2797	33.5913	80.837	−51.463	−30.886
MeCN	0.1703	0.2812	37.8509	83.913	−50.432	−31.026
DMF	0.2766	0.3011	66.8053	102.408	−4.974	−15.536
F	0.3273	0.3131	82.5557	108.820	15.127	−3.368
NMF	0.3310	0.3127	79.2671	107.355	16.900	−2.834
DMSO	0.3171	0.3117	86.2748	110.430	10.915	−5.788
n-BuOH	0.3173	0.3247	91.6472	112.676	4.552	−0.820
i-BuOH	0.1418	0.2229	24.8461	72.925	−39.403	−47.535
t-BuOH	0.1879	0.2704	31.1935	78.671	−33.602	−30.548

[a] F = formamide; NMF = N-methylformamide; DMF = N,N-dimethylformamide.

Table 5.8 Values of ΔE_{ab}^* and ΔH_{ab}^* differences between colours of CoBr$_2$ solutions in a solvent and water ($c = 0.01$ M)

Solvent[a]	ΔH_{ab}^*	ΔE_{ab}^*
Me$_2$CO	30.6515	71.8643
MeCN	30.4224	69.6815
i-BuOH	29.9625	76.7213
t-BuOH	18.1449	60.5021
DMF	14.2278	21.8877
DMSO	5.2766	10.3845
F	3.9087	13.0572
NMF	3.6585	15.0094
MeOH	2.6631	4.5210
n-PrOH	1.9654	2.2524

[a] F = formamide; NMF = N-methylformamide; DMF = N,N-dimethylformamide.

Tables 5.6 and 5.7 show clear differences in the chromaticity of solutions in different solvents. These differences are shown graphically in the b^*−a^* diagrams presented as Figures 5.8 and 5.9.

On the basis of the above data and on the assumption that Co^{2+}aq has an octahedral structure in water – an assumption which is soundly based on the interpretation of absorption spectra and on the demonstration of a hydration number of six by NMR spectroscopy [18] – we can establish the following direction of ΔE_{ab}^* changes for the CoBr$_2$-solvent system (as compared with water and for a given concentration):

i-PrOH > A > AN > t-BuOH > DMF > NMF > F > DMSO > MeOH > n-PrOH

In solvents towards the end of the series the Co^{2+} solvento-ions are generally octahedral, whereas in solvents towards the beginning of the series either tetrahedral forms predominate or both octahedral and tetrahedral forms occur.

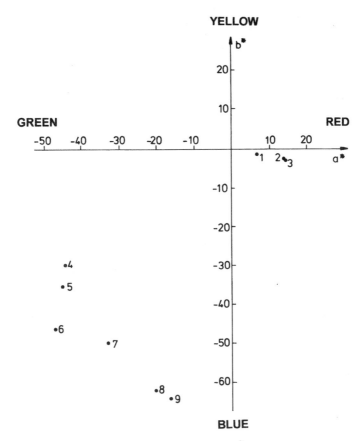

Figure 5.8 CIELAB chromaticity diagram of $0.01\,mol\,dm^{-3}$ CoCl$_2$ solutions in 1 – methanol, 2 – N-methylformamide, 3 – formamide, 4 – dimethyl sulphoxide, 5 – dimethyl formamide, 6 – acetonitrile, 7 – acetone, 8 – iso-propanol, and 9 – t-butyl alcohol [16].

Chromaticity may be used for the purpose of objective (psychophysical) evaluation of solution colour for any coloured complex. Such an indicator serves its function even if the perceived colours in different systems differ only slightly, but the differences show up in the measurements of the values characterising the chromaticity of the solutions. One such value is the dominant, or complementary, wavelength, formulated in Chapter 1. In blue CoCl$_2$ solutions, the dominant wavelength for the solvents examined ranges from 479 to 487 nm. For solutions in methanol, in formamide, and in N-methylformamide the complementary wavelength must be determined, as chromaticity coordinates position these colours within the purple triangle.

An important practical role is played by the value of the so-called hue angle (as defined earlier in this book, in Section 1.5.2). For solutions of CoCl$_2$ clear differences are found between two groups of solvents. All blue solutions show a hue angle between 215 and 255°, the others have a hue angle of about 350°.

Absorption spectra of 0.01 M CoBr$_2$ solutions in solvents conducive to the formation of tetrahedral solvates show huge differences in colour, expressed as ΔE_{ab}^*,

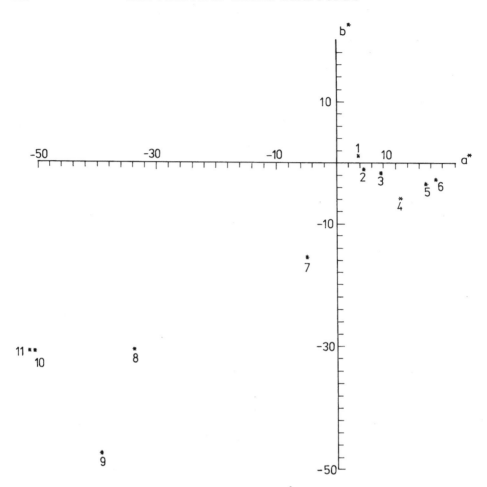

Figure 5.9 CIELAB chromaticity diagram of $0.01 \, mol \, dm^{-3}$ $CoBr_2$ solutions in 1 – water, 2 – n-propanol, 3 – methanol, 4 – dimethyl sulphoxide, 5 – formamide, 6 – N-methylformamide, 7 – dimethyl formamide, 8 – t-butyl alcohol, 9 – iso-propanol, 10 – acetonitrile, and 11 – acetone [17].

from comparable aqueous solutions (Table 5.6) in which the cobalt is octahedrally coordinated (the band at approximately $20\,000 \, cm^{-1}$). However, these bands differ principally in intensity, which in practice corresponds to the proportion of the octahedral form and which causes an effective colour shift expressed as a change of chromaticity (see Figures 5.10, 5.11, and 5.12 for $CoCl_2$ and $CoBr_2$).

Chromaticity coordinates have been used in soil analysis to determine iron content and consequently to assess the suitability of the soil. The Kubelka-Munk method has been applied to the examination of reflectance spectra of various mixtures of synthetic iron oxides, hematite and goethite, and samples of iron-free soil [19]. Without going into details of interest only to specialists in the field, it should be noted that the authors believe that the Kubelka-Munk theory can be used to predict the colour of mixtures of soil with synthetic iron oxides (hematite and goethite).

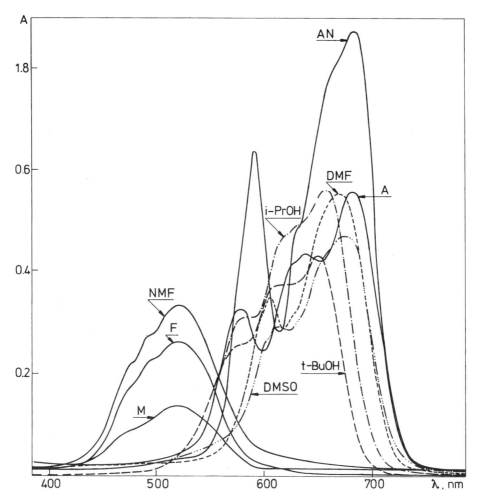

Figure 5.10 Absorption spectra of CoCl$_2$ solutions in organic solvents (A = acetone, AN = acetonitrile, F = formamide, M = methanol) [16].

Moreover, theoretical soil colour indices could be formulated, correlated with the content of these oxides and mineralogical parameters. The authors used this approach for the analytical determination of content and type of iron oxides; CIE, CIELAB, and Munsell systems were used. Figure 5.13 presents the calculated CIELAB chromaticity coordinates for mixtures with various percentage contents of hematite (Hm) and goethite (Gt). The form of the dependence is the same as that in the x, y, Y system for simulated spectra in the $Cr^{3+} - H_2O - NH_3$ system described in Chapter 3.

Interesting studies have also been made on computing the colour of soil from reflectance spectra, and of its dependence on organic content [20, 21]. Such dependence was found as early as the 1920s, but the lack of standards at that time

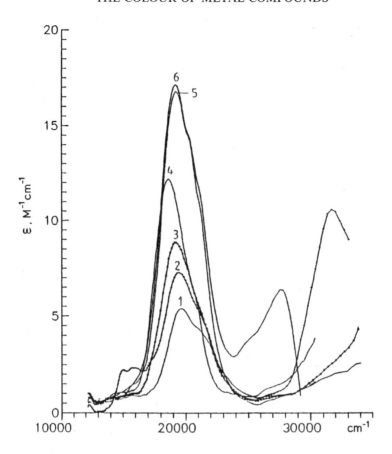

Figure 5.11 Absorption spectra of CoBr$_2$ solutions in organic solvents: 1 – water, 2 – n-propanol, 3 – methanol, 4 – dimethyl sulphoxide, 5 – formamide, 6 – N-methylformamide (these solvent numbers correspond to those in Figure 5.9) [17].

made it impossible to give a precise formulation of the dependence. It is now apparent that there is a linear correlation ($r^2 > 0.9$) between the Munsell value and the content of organic matter in the soil. Based on twelve soil samples, the following equations were given [21]:

$$y = 6.33 - 0.0511x, \quad \text{for dry samples } (r^2 = 0.92);$$

$$y = 4.38 - 0.0523x, \quad \text{for moist samples } (r^2 = 0.94).$$

In these equations y denotes the Munsell value and x the content of organic matter, in grams per kilogram.

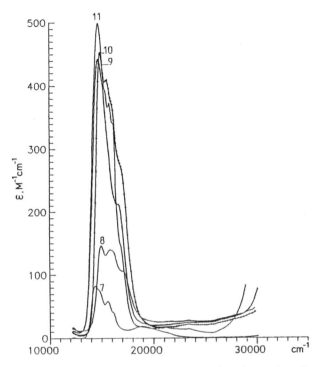

Figure 5.12 Absorption spectra of CoBr$_2$ solutions in organic solvents: 7 – dimethyl formamide, 8 – t-butyl alcohol, 9 – iso-propanol, 10 – acetonitrile, and 11 – acetone (these solvent numbers correspond to those in Figure 5.9) [17].

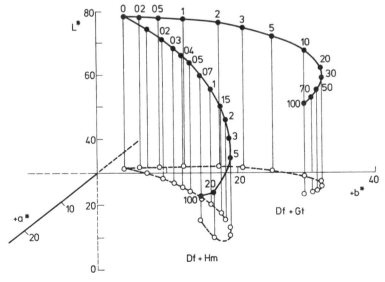

Figure 5.13 Calculated positions (•) and their projections on the a^*-b^* plane (o) of hematite-deferrated soil (Hm + Df) and goethite-deferrated soil (Gt + Df) in the CIE ($L^*a^*b^*$) colour space L^* = lightness, a^* = redness, b^* = yellowness). The numbers refer to the percent hematite or goethite.

REFERENCES

[1]   Aleksiejew, W., 1955, *Chemia analityczna*, PWN, Warsaw.

[2]   Allen, T.L., 1958, *Analyt. Chem.*, **30**, 447.

[3]   Marczenko, Z., 1979, *Spektrofotometryczne oznaczanie pierwiastków, 3rd edn.*, PWN, Warsaw; 1976, *Spectrophotometric Determination of Elements*, Ellis Horwood, Chichester.

[4]   Barbosa, J., Barron, D. and Bosch, E., 1987, *Analyst*, **112**, 1717.

[5]   Kolthoff, I.M. and Elving, P.J., 1959, *Treatise on Analytical Chemistry, Part I, Vol. 1*, Wiley, New York; 1979, *Part I, Vol. 2, 2nd edn.*, Wiley, New York.

[6]   Krishna Prasad, K.M.M., Raheem, S., Vijayalekshmi, P. and Kamala Sastri, C., 1996, *Talanta*, **43**, 1187.

[7]   Silver, G.L., 1967, *Talanta*, **14**, 637.

[8]   Barbosa, J. and Bosch, C.M., 1991, *Talanta*, **38**, 1297.

[9]   Reilley, C.N., Flaschka, H.A., Laurent, S. and Laurent, B., 1960, *Analyt. Chem.*, **32**, 1218.

[10] Roses, M., 1988, *Anal. Chim. Acta*, **204**, 311.

[11] Gillard, R.D. and Sutton, H.M., 1970, *J. Chem. Soc. (A)*, 1309, 2172, 2175.

[12] Soukup, R.W. and Schmid, R., 1985, *J. Chem. Educ.*, **62**, 459.

[13] Sone, K. and Fukuda, Y., 1972, *Bull. Chem. Soc. Japan*, **45**, 465; Sone, K. and Fukuda, Y., 1982, *Stud. Phys. Theor. Chem.*, **27**, 251.

[14] Sommer, L., 1958, *Z. Anal. Chem.*, **164**, 299.

[15] Tłaczała, T. and Bartecki, A., 1997, *Monats. Chem.*, **128**, 225.

[16] Bartecki, A. Tłaczała, T. and Raczko, M., 1991, *Spectr. Lett.*, **24**, 559.

[17] Bartecki, A. and Tłaczała, T., 1993, *Spectr. Lett.*, **26**, 809.

[18] Matwiyoff, N.A. and Darley, P.R., 1968, *J. Phys. Chem.*, **72**, 2659.

[19] Barron, V. and Torrent, J. 1986, *J. Soil Sci.*, **37**, 499.

[20] Fernandez, R.N. and Schulze, D.G., 1987, *Soil Sci. Soc. Amer. J.*, **51**, 1277.

[21] Fernandez, R.N., Schulze, D.G., Coffin, D.L. and Van Scoyoc, G.E., 1988, *Soil Sci. Soc. Amer. J.*, **52**, 1023.

6. COLOUR CENTRES

6.1 FORMATION AND SPECTROSCOPIC PROPERTIES

The problem of colour centres, which emerged at least a hundred years ago, arose from the observation of the occurrence of some coloured specimens of minerals that are normally colourless. A classic example is provided by natural rock salt crystals, which show yellow colouring and absorb light in the blue range of the spectrum. In this book we can only do justice to the most important aspects of this issue, but in the literature, both earlier and more recent, one can find excellent monographs and numerous articles (e.g. [1–5]).

The causes of the occurrence of colour centres are generally related to transitions between energy bands. In a broader, though slightly loose, approach, one can also apply ligand field theory to electronic transitions in the transition metal impurities in the given crystal. However, the classical theories of the mechanism of colour centre formation point to certain specific features distinguishing them from the mechanisms of colour formation in transition metal compounds described in the preceding chapters.

The principal type of objects where such centres occur are halides of the alkali metals. Their crystallographic structures are well known and mostly belong to two types: the face-centred structure of NaCl and the body-centred structure of CsCl. These structures contain lattice defects, so-called Frenkel and Schottky defects. Figure 6.1 presents in a schematic way the formation of those defects [4].

In the Frenkel defect, there is a shift of an ion from its normal position at a crystal node to another position where such ions usually do not occur, to an interstitial site. A pair of defects is formed: a vacancy and an ion in this new position. In the Schottky defect, there are two vacancies but no ion in a different position. The vacancies may occur as a result of the movement of two ions with opposite charges to the surface of the crystal. Even though the mechanism of the formation of these two defects is different, they both preserve the electrical neutrality of a crystal.

Colour centre formation is thus related to crystal lattice defects, which occur in practically all crystals at approximately 0.01% of the content, but the defects alone

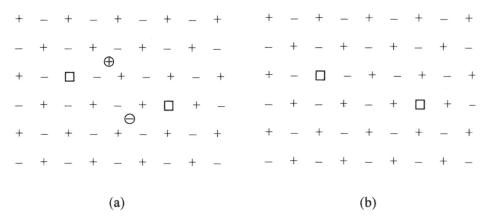

(a)   (b)

Figure 6.1 Frenkel (a) and Schottky (b) defects (schematic) [4].

do not lead to the occurrence of colour centres. Various experimental methods of colour centre formation are known. The main ones include:

a) additive colour centre formation using an excess of metal;

b) the same process with an excess of the anion, that is, in practice, an excess of the halogen (in alkali metal halide lattices);

c) galvanic colouring;

d) colour centre formation as a result of ionising radiation.

The principal mechanism of colour centre formation consists in the trapping of electrons or holes by crystal lattice defects. This leads to the formation of various centres, whose absorption spectra show the presence of numerous bands corresponding to different kinds of defects and their combinations. Absorption bands are usually denoted by commonly adopted symbols, such as F, F', R, V and others. Figure 6.2 presents examples of the absorption spectra of the F colour centres in alkali metal chlorides [4].

One of the most frequently and most thoroughly studied centres is the F centre. On the most straightforward interpretation, the mechanism of the formation of such

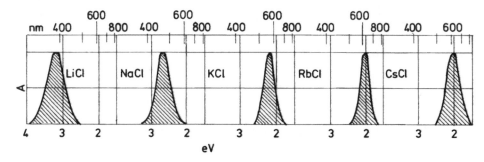

Figure 6.2 Absorption spectra of F colour centres in alkali halides [4].

centres consists of two stages. In the case of an alkali metal halide, if excess metal is applied, the following reactions take place:

$$M \longrightarrow M^+ + e^-$$

with an additional formation of crystals on the surface by a metal ion and a halogen ion from the crystal lattice,

$$M^+ + X^- \longrightarrow MX,$$

where X^- denotes a halide ion.

An excess electron knocked out of a metal atom is trapped by an ionic vacancy, thus forming an F centre. According to band theory, the excitation of an electron occurs, causing its transition from the valency band to the conduction band. From the fact that the energy gap in alkali metal halides is approximately 5–12 eV (i.e. about 250–100 nm) it can be inferred that even ultraviolet radiation may be too weak to knock an electron out. An electron in the conduction band is attracted by a halide anion vacancy, which is positively charged. While being in the vacancy, the electron is trapped by the surrounding positively charged cations. As a result, new energy levels are created between the vacancies and the conduction band, and the transitions occurring cause the formation of colour centres. It is worth noting that some authors (e.g. [6]) state that colour centre formation can also be explained by the formalism of ligand field theory. A vacancy is treated as an atom, while the electron has excited energy levels as a result of interactions with ligands.

The colour of a colour centre (Figure 6.2) corresponds to the position of the absorption band. Various empirical formulae are known for the determination of its energy (or wavelength). For crystals with the NaCl structure, it has been found [4] that the wavelength expressed in Å (at room temperature) is given by the following relationship:

$$\lambda_{max} = 600 \, d^2,$$

where d denotes the lattice constant of the alkali metal halide (in Å).

Another formula was given by Mollwo and Yvey (from [1]):

$$E_a = \frac{17.4}{d^{1.84}},$$

where E_a is expressed in eV, and d in Å.

Table 6.1 contains data concerning F centres for different alkali metal halides [7]. It has been found that in a typical strongly coloured crystal there is one colour centre per about 10 000 halide ions.

It follows from the above data that for a given halogen ion, for instance for Cl^-, the energy E_a decreases with the atomic number of the alkali metal, and the colour changes gradually from yellow green through brown yellow and purple to blue green, that is clockwise in the chromaticity diagram. Since there is one characteristic absorption band here with a purely Gaussian shape (as is assumed in the literature),

Table 6.1 Absorption band energies and wavelengths,[a] and colours, of F centres in some alkali metal halides

Halide:	F			Cl			Br		
Metal	Energy (eV)	λ (nm)	Colour	Energy (eV)	λ (nm)	Colour	Energy (eV)	λ (nm)	Colour
Li	5.083	239	colourless	3.2	387	yellow green	2.7	459	yellow brown
Na	3.707	334	colourless	2.746	451	yellow brown	2.345	529	purple
K	2.873	431	yellow brown	2.295	540	violet	2.059	602	blue green
Rb	2.409	514	purple[b]	2.034	609	blue green	1.852	669	blue green

[a] Energies are from ref. [3]; λ values have been calculated from $λ = 1239.9/eV$. [b] This colour was not given in ref. [3].

such systematic changes can be predicted. Gradual changes of the energy E_a correspond to changes of ionisation potential, which occur in accordance with the first stage described above.

Absorption bands corresponding to F centres show a clear dependence on temperature, as can be seen in Figure 6.3 [4]. The dependence is expressed by the following equation:

$$W = A \coth (hv_g/2kT)^{1/2},$$

where A and v_g are constants, h and k are respectively the Planck and the Boltzmann constants, and W denotes the value of the total halfwidth of the band.

The oscillator strengths of the absorption bands of F centres have also been calculated on the assumption that there is no interaction between them. In such a case, the experimentally measured area under the spectrum curve is proportional to the concentration of centres.

As has already been mentioned, there may be other centres, apart from F centres, in alkali metal halide crystals. Figure 6.4 presents the absorption spectra (bands) corresponding to such centres [7].

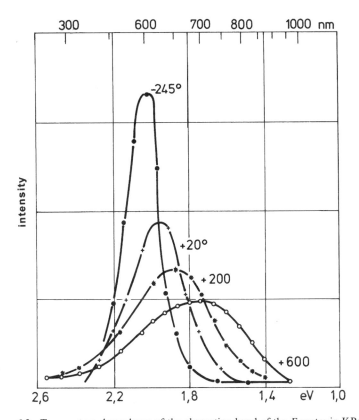

Figure 6.3 Temperature dependence of the absorption band of the F centre in KBr [4].

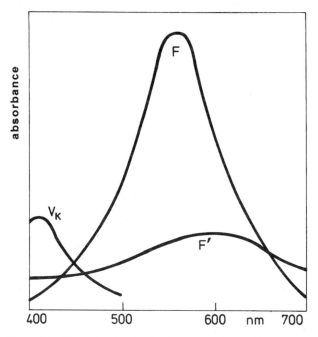

Figure 6.4 Absorption bands of colour centres in a KCl crystal [7].

The symbol F' or F^- is used to denote a colour centre which is formed if a crystal is exposed to radiation corresponding to the wavelength of F at low temperatures. As a result the intensity of the F band decreases and an asymmetrical and very broad F' band appears. It is assumed that only some F centres change into F' centres, so there may be coupling between the two types of absorption band. F' centres are regarded as F centres that have trapped another electron.

Prolonged exposure to Roentgen radiation at room temperature also brings about the formation of other colour centres and absorption bands shifted bathochromically as compared with F bands. The underlying mechanism of their formation is the same as for F centres.

Figure 6.4 also shows an absorption spectrum denoted with the symbol V_k. In contrast to the cases discussed so far, this spectrum characterises colour centres formed as a result of trapping holes, not electrons. Similarly, H-centres involve hole trapping at an interstitial halide ion. These hole-trapped centres are stable only at low temperatures. A range of centres may be seen in the schematic representations shown in Figure 6.5 [3] and elsewhere [5].

As has been mentioned, the above centres may also be formed by an additive method, that is in the case of alkali metal halides by means of an excess of the halogen. Figure 6.6 shows the absorption spectrum of KI with excess I_2. The crystal was heated up to the temperature of 600 °C at the pressure of 10 atm. of I_2, and subsequently cooled down at room temperature in CCl_4 [4].

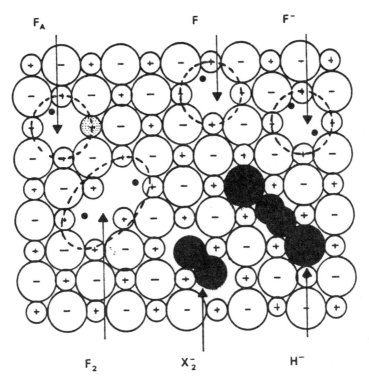

Figure 6.5   Models of the main types of colour centre in alkali halides [3].

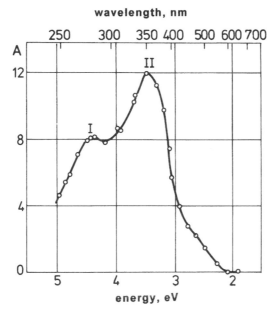

Figure 6.6   Absorption spectrum of the colour centres in KI containing an excess of I_2 [4].

6.2   COLOUR CENTRES AND IMPURITIES IN THE MATRIX LATTICE (ALKALI METAL HALIDES AND OTHERS)

In the above brief treatment of colour centre formation in alkali metal halides we considered pure crystals containing no added impurities, or at least containing such a low level of impurity that it could be disregarded. However, the introduction of minute quantities of other compounds (alias impurities) may have a significant effect on the properties of colour centres. Three factors are significant [4]:

1. The characteristics and the role of the absorption bands of the impurities themselves.

2. Changes of the absorption bands of colour centres caused by impurities.

3. Effect of impurities on colour formation.

The effects of impurities are believed [4] to arise from one or more of the following contributory causes:

a) The ions of the impurities may form absorption or luminescence centres themselves.

b) Due to differences between the size of the impurity ions and those of the matrix crystal, there may be differences of lattice constants and some local stresses may occur, resulting in changes of transition energy, that is the position and width of the absorption band (and hence in changes of colour).

c) Differences between the charge of the impurity ions and those of the matrix crystal may bring about a change in the concentration of positive and negative vacancies, as well as other lattice defects, thus causing a strengthening or a weakening of the coloration of the crystal.

d) Differences between charges may also lead to the formation of 'complexes' of impurities with vacancies, which trap electrons or holes and thus cause the formation of new types of colour centres which do not occur in pure crystals.

e) Because of differences in the ionisation potentials or electron affinity between the impurity ions and those of the crystal, electrons and holes may sometimes be trapped more easily by the impurities than by the lattice defects in the matrix crystal, which leads to the occurrence of new colour centres.

One important type of systems which has long been the object of interest, comprises crystals (mostly of alkali metal halides but also of other matrices) doped with cations of bivalent metals, e.g. Ca^{2+}, Sr^{2+}, Cd^{2+}, Pb^{2+}, and Mn^{2+}, as well as +1 cations, e.g. Ag^+, Cu^+, or Tl^+.

In the case of bivalent cations, not all of them cause the formation of colour centres affecting the colouration of the crystal. Thus for instance, in the case of NaCl:Pb^{2+}, absorption bands appear at 193, 273, and 290 nm, while emission bands excited by radiation with a wavelength of 275 nm have positions at 318, 384, and 453 nm [4].

More important here are crystals doped with ions of the heavier alkaline earth metals. Thermal and optical methods applied to such crystals result in the formation of numerous colour centres characteristic of these impurities. Four types of such colour centres have been studied, denoted by the symbol Z_{1-4}. Additionally there are F centres in such systems. Table 6.2 gives the positions of absorption bands in four systems [4].

Table 6.2 Absorption band maxima due to Z and F colour centres in some alkali metal halides doped with M^{2+} alkaline earth ions [4]

System	wavelength of maxima (nm)				
	F	Z_1	Z_2	Z_3	Z_4
KCl:Sr	540	595	635	505	860
KCl:Ba	540	600	650		
KCl:Ca	540	590	610		
NaCl:Sr	450	505	512		

The dependence of the position of absorption bands on the kind of impurity suggests that the impurity ion is a part of the colour centre. Figure 6.7 shows an example of F and Z centres in the KCl:Sr^{2+} system [4].

An interesting question, particularly in connection with the main ideas of this book, is the role of Mn^{2+} as a d-electronic element. The lowest-energy optical excitation (in the gaseous phase) occurs between the atomic levels ^6S–^6G (for a high-spin d^5 configuration). The concentration of Mn^{2+} ions occurring as impurities in alkali metal halide crystals is usually very low, while the molar absorption coefficients are in the order of 10^{-2}–10^{-3}, which practically precludes the observation of Mn^{2+} absorption bands. Direct excitation of the energy levels of the impurity ions is difficult too. It has been found, however, that if in addition to Mn^{2+} other ions are also introduced to NaCl or KCl crystals, such as Pb^{2+}, it is possible to obtain optical excitation of, and subsequently emission from, Mn^{2+} ions. In such cases there is absorption by the Pb^{2+} ion, energy transfer to Mn^{2+}, and emission from the energy levels of that ion. The closer these ions are to each other in the crystal, the easier the emission is.

In the NaCl:Mn^{2+} system, Roentgen radiation causes the formation of F centres, and subsequently optical excitation brings about luminescent emission characteristic of the Mn^{2+} ion.

A large number of studies have been conducted on impurities in the form of univalent heavy metal ions, such as Ag^+, Cu^+, and Tl^+. These ions cause the occurrence of new absorption bands depending on the kind of ion and matrix. In the majority of such systems, the additional absorption bands lie outside the visible range, but emission bands occur also within the visible range. For instance in the KCl:Tl^+ and KI:Tl^+ systems their positions are respectively 475 and 415 nm.

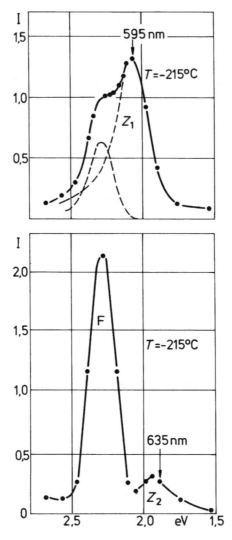

Figure 6.7 Absorption spectra of F and Z colour centres in KCl:Sr [4].

REFERENCES

[1] Farge, Y. and Fontana, M.P., 1979, *Electronic and Vibrational Properties of Point Defects in Ionic Crystals*, North-Holland, Amsterdam.

[2] Henderson, B., 1987, *Optical Spectroscopy of Color Centres in Ionic Crystals*, in *Spectroscopy of Solid-State Laser-Type Materials*, ed. Di Bartolo, B., Plenum, New York.

[3] Henderson, B. and Imbusch, G.F., 1989, *Optical Spectroscopy of Inorganic Solids*, Clarendon, Oxford.

[4]   Schulman, J.H. and Compton, W.D., 1962, *Color Centres in Solids*, Pergamon, New York.

[5]   Brown, F.C., 1997, *Color Centers*, in *McGraw-Hill Encyclopaedia of Science & Technology*, 8th edn., McGraw-Hill, New York, Vol. 4, p. 180.

[6]   Nassau, K., 1983, *The Physics and Chemistry of Color*, Wiley, New York.

[7]   Levy, P.W., 1981, *Color Centers*, in *Encyclopaedia of Physics*, ed. Lerner and Trigg, Addison-Wesley, Reading, Mass., p. 132.

7. COLOUR AND ELECTRONIC SPECTRA OF MINERALS, PIGMENTS, AND GEMSTONES CONTAINING TRANSITION METALS

7.1 GENERAL REMARKS

Numerous minerals are a natural source of d- and f-block transition metals, particularly the former. As regards their colour, one should distinguish between allochromatic and idiochromatic minerals [1]. In the former case transition metals, especially from the $3d$ series, occur as impurities. Despite their small or very small concentrations they often impart a colouration characteristic of the particular element in the appropriate oxidation state and ligand environment (symmetry). An example of such a mineral is ruby, in which Cr^{3+} ions replace Al^{3+}, to give the mineral its characteristic colour. In the case of idiochromatic colours, transition metal compounds are the principal or dominant component determining the colour of the mineral. There may at the same time be other elements in minute quantities affecting the optical spectra and the colour of minerals. Tables 7.1 and 7.2 contain some examples of both types of minerals. As can be seen from both these tables, $3d$-electronic elements occurring in different stable oxidation states play a crucial role here. However, the problem of the colour of minerals in both the groups is extremely complex. This book presents just a brief account of the issues involved.

7.2 COLOUR AND ELECTRONIC SPECTRA OF MINERALS

These topics are discussed in many papers and monographs, of which the present authors have found five particularly useful [2, 3, 4, 5, 6]. We emphasise again here that neither idiochromatic nor allochromatic colours can necessarily be ascribed to a single transition metal or a single type of electronic transition. In elements of the $3d$ transition series, which are being considered here, ligand field transitions are the most commonly encountered, but other kinds of transitions also occur.

As we know from our earlier discussion, the colour of a transition metal compound depends not only on the composition of its immediate surroundings – the

Table 7.1 Selected examples of minerals (m), gemstones (g), and pigments (p) whose idiochromatic colours arise from d-d transitions[a]

Element	Oxidation state	Compound	Formula	Colour
Cr	3+	chrome oxide green (p)	Cr_2O_3	green
		uvarovite (m, g)	$Ca_3Cr_2Si_3O_{12}$	green
		potassium chrome alum	$KCr(SO_4)_2.12H_2O$	purple (violet)
Mn	2+	rhodonite (m, g)	$MnSiO_3$	pink
		rhodochrosite (m, g)[c]	$MnCO_3$	red
		spessartine (m, g)	$Mn_3Al_2(SiO_4)_3$	red[b]
	3+	manganite(m)	$MnO(OH)$	brown
Fe	2+	siderite (m, g)	$FeCO_3$	brown yellow
		fayalite (m)	Fe_2SiO_4	brown green
	3+	hematite (m, p)	Fe_2O_3	red brown
		goethite (m)	$FeO(OH)$	yellow, yellow red
Co	2+	spherocobaltite (m)	$CoCO_3$	pink
Ni	2+	bunsenite (m)	NiO	green
Cu	1+	cuprite (m, g)	Cu_2O	red
	2+	malachite (m, g)[c]	$Cu_2(CO_3)(OH)_2$	green
		turquoise (g)	$CuAl_6(OH)_8(PO_4)_4.4H_2O$	blue, green
		dioptase (m, g)	$Cu_6Si_6O_{18}.6H_2O$	green
		chrysocolla (m)[c]	$Cu_2Si_2O_5(OH)_2.2H_2O$	blue, blue green
Mo	4+	wulfenite (m, g)[c]	$PbMoO_4$	yellow, orange

[a] The *Encyclopaedia of Minerals and Gemstones*, ed. M. O'Donoghue, Orbis, London, 1976, is one of a number of comprehensive and copiously illustrated volumes providing detailed information in this area. [b] From yellow red to brown red (also orange). [c] Colour pictures of these may be found as Figures C12 to C15.

Table 7.2 Examples of allochromatic colours caused by ligand field transitions in $3d$ ions

Transition metal impurity	Substance	Colour
Chromium	ruby (Al_2O_3), natural or synthetic	pink[a], red[a]
	emerald	green
	sapphire	blue, yellow, green[a]
	topaz ($Al_2F_2SiO_4$)	pink red, yellow, green, blue, brown
	alexandrite	green + red (dichroism)
Manganese	tourmaline (boron aluminium silicate)	green, pink, red
	andalusite	green, yellow, brown, red (rarely)
Vanadium	vanadium grossular garnet	green, pink, yellow, brown
	vanadium emerald	green
Cobalt	spinel (synthetic) ($MgAl_2O_4$)	blue

[a] Colour and spectra of these specimens are discussed in detail in the text.

nature of ligand donor atoms – but also on the ion-ligand distance. This has a direct effect on the value of Dq. Thus, for instance, Cr(III) causes a green coloration in Cr_2O_3, but in potassium chromium alum, $KCr(SO_4)_2.12H_2O$, it produces a purple (violet) coloration; other examples are given in Chapter 3. The presence of Mn(II) in $MnCO_3$ (rhodochrosite) is the source of this gemstone's red colour, while in the mineral MnO it produces a green colour, in the garnet known as spessartine, $(Mn_3Al_2Si_3O_{12})$ it gives an orange colour, and in potassium manganese alum violet. Of course the concentration of the chromogenic element is an important factor, often determining not only the hue but also the purity of the colour. We shall return to this matter shortly. The presence of other elements in a mineral may also have a significant effect.

The handful of examples presented above show that colour alone cannot be the basis for the identification of minerals. Undoubtedly electronic spectra, often obtained as absorption spectra, of minerals are more important in this respect. The technical aspects of such measurements are not simple [2], but often satisfactory spectra can be obtained and wavelengths (wavenumbers) of maximum and minimum absorption and their respective molar absorption coefficients determined. Interpretation, however, is difficult. In addition to ligand field transitions there may also be MLCT or LMCT transitions, or transitions between ions of the same metal in different oxidation states or between ions of different metals (MMCT or IVCT transitions). It should also be noted that most authors do not use computer analysis of the spectral contour. Without such treatment it is extremely difficult to analyse electronic transitions unequivocally or to determine precisely spectral parameters and Dq values, let alone the Racah parameters B and C. Nevertheless, the results presented in some publications do look convincing; they can usefully be compared with measurements on appropriate pure compounds.

If we consider the question of spectra and colour from the point of view of the main chromogenic element, we can conclude that iron and manganese are the most important, with chromium also playing an important role in gemstones.

7.2.1 Electronic spectra and colour related to the occurrence of chromium ions

We shall discuss first the question of the allochromatic colour of ruby and of alexandrite, which has been presented in a very lucid manner elsewhere [1]. The electronic spectrum of ruby, with a simplified Tanabe-Sugano diagram in an octahedral field and a diagram of the electronic transitions, are shown in Figure 7.1. As can be seen, two absorption bands are related to transitions from the $^4A_{2g}$ level to the $^4T_{2g}$ level with an energy of 2.23 eV $\approx$ 18 000 cm^{-1} (555 nm) and to the $^4T_{2g}$ level with an energy of approximately 3.1 eV $\approx$ 25 000 cm^{-1} (400 nm). The energy of the first transition determines the value of the $10Dq$ parameter. As a result of strong absorption in the violet region and almost as strong in the yellow green region, together with strong transmittance in the red region and somewhat weaker in the blue region, ruby exhibits its characteristic colour. It is additionally strengthened by red fluorescence, or rather phosphorescence as this is a transition between states of different multiplicity, with an energy of approximately 1.79 eV $\approx$ 14 436 cm^{-1} (692 nm).

A dramatic change of colour is observed if Cr^{3+} occurs as an impurity in emerald with the composition $Be_3Al_2Si_6O_{18}$. In emerald, $10Dq$ is slightly reduced, to 2.05 eV $= 16 535$ cm^{-1}, while the other transition has an energy of 2.8 eV $= 22 585$ cm^{-1} [7]. Both the change of transition energies and a slight change of the spectrum contour cause a shift of absorption from the yellow green region to yellow

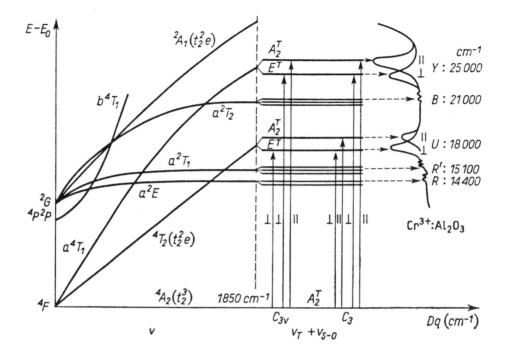

Figure 7.1 Energy level diagram and absorption spectrum parameters for ruby. $\perp$ and $\|$ denote transition polarisation; $\nu - O_h$ symmetry; $\nu_T - C_{3\nu}$ symmetry (trigonal); ν_{S-O} spin-orbit coupling [8].

red with sustained strong transmittance in the blue green range. As a result emerald has the characteristic green colour.

Let us recall in this connection that we have shown above (in Chapter 3) in simulation studies the significance of even slight changes of Dq for the CIE colour in Cr^{3+} compounds. For $Dq = 1790\,cm^{-1}$, simulation showed red, while for $1640\,cm^{-1}$ it indicated green (for 1 M concentration).

Further, it turns out that there is a mineral, alexandrite, whose Dq value is between that of ruby and that of emerald. The mineral is beryllium aluminate, $BeAl_2O_4$, with Cr^{3+} ions as impurities. Without Cr^{3+} ions the mineral is colourless and is known as chrysoberyl. In alexandrite, the Dq value is 2.17 eV (17 503 cm^{-1}), intermediate between 2.23 eV for ruby and 2.05 eV for emerald. What is interesting is the colour of this mineral. Whereas in daylight, rich in blue-range light, the mineral is blue green, resembling emerald to a certain extent, in light that is rich in red, the colour assumes a strong red shade, thus resembling ruby. This phenomenon is called the 'alexandrite effect' and also occurs in some other gemstones.

In minerals or gemstones with crystallographic structure other than cubic, the phenomenon of dichroism, or more generally pleochroism, occurs. It is due to the fact that different wavelengths of polarised light are absorbed in varying degrees along different crystallographic axes. As a result, different colours occur when a crystal is viewed in transmitted light. On this basis, it can also be concluded that pleochroism is a proof that a given ion is in a particular environment symmetry lower than cubic. One of the main types of pleochroism is dichroism, for instance in ruby, which has only one characteristic crystallographic axis; if there are two such axes, trichroism occurs.

In the case of ruby, the situation is that Cr^{3+} ions occur in the corundum structure in distorted trigonal crystal field symmetry. As a result, both the quartets undergo a slight splitting, which can be seen in Figure 7.1 [8]. The colour of ruby caused by the absorption of polarised light parallel to its main axis is orange red, whereas absorption perpendicular to this axis results in red purple colour [1].

Alexandrite has two optical axes and hence manifests trichroism, resulting in the occurrence of red purple, orange, and green colours. It must be clearly said that pleochroism is connected with deformation of cubic symmetry, which causes a splitting of energy levels. The polarised spectrum of alexandrite is shown in Figure 7.2 [4].

The colour of minerals, or gemstones, is strongly dependent on the content of the chromogenic element. Figure 7.3 [9] shows the dependence of ligand field strength (in eV) on the amount of Cr_2O_3 (in mole %).

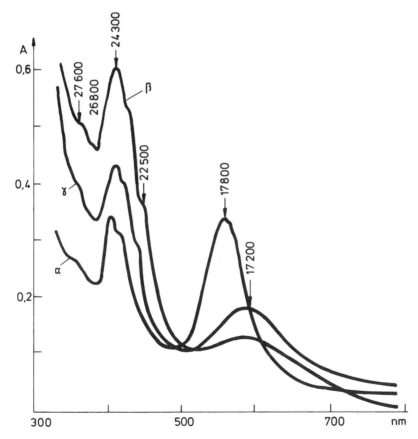

Figure 7.2 Polarized absorption spectrum of an alexandrite specimen (from the Urals). α, β, γ denote three polarized components; data are in wavenumbers [4].

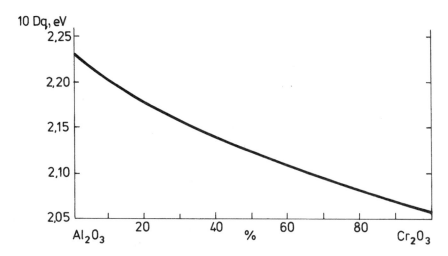

Figure 7.3 The dependence of $10Dq$ in the system colourless Al_2O_3 (sapphire) $+Cr_2O_3$. Colour changes are discussed in the text [9].

Pure Al_2O_3 (corundum) is colourless, and with the amount of Cr_2O_3 gradually increasing, first pink and then red (ruby) colour appears. When the Cr_2O_3 content reaches about 25%, the colour becomes grey and subsequently changes gradually into more and more intense green, until it reaches the green colour of pure Cr_2O_3. The $10Dq$ parameter in this case is approximately 2.06 eV ($16615\,cm^{-1}$). For $10Dq = 16\,700\,cm^{-1}$, assumed in simulation calculations, the chromaticity coordinates obtained from calculations were $x = 0.0895$, $y = 0.6300$, and $Y = 0.0581$ (for 1 M concentration), corresponding to the green colour.

Characterising sapphires as minerals that owe their colour to the presence of Cr^{3+}, it should be stated that the occurrence of other transition elements and other types of electronic transitions, in addition to ligand field transitions, may lead to the occurrence of various colours in some minerals from different deposits. Corundums from Nepal have been found to be pink, violet red, or violet [10]. Figure 7.4 shows the absorption spectra of six such minerals. The visible absorption bands are due to the presence of Cr^{3+} but, according to the authors, the red violet and violet sapphires also show additional absorption bands in the red region, related to Fe^{2+}/Ti^{4+} charge-transfer transitions. This kind of transition is also found in other minerals.

The same authors also investigated blue and yellow sapphires from Nigeria. The relevant spectra of the former are given in Figure 7.5 [10]. As can easily be seen, the absorption spectra of Cr^{3+} ions do not show their typical nature in this case. More important here are Fe^{3+} ions and Fe^{2+}/Ti^{4+} and Fe^{2+}/Fe^{3+} pairs. Thus, blue colour is not necessarily related to the presence of Cr^{3+}, which is attested by the absorption spectra of sapphires which do not contain that ion (Figure 7.6). In addition to crystal field transitions, there are CT transitions in pairs of ions here, but their spectroscopic interpretation must be very circumspect.

The main component causing the occurrence of colour in both natural and synthetic rubies consists of Cr(III) ions, with the amount of Cr_2O_3 in very strongly

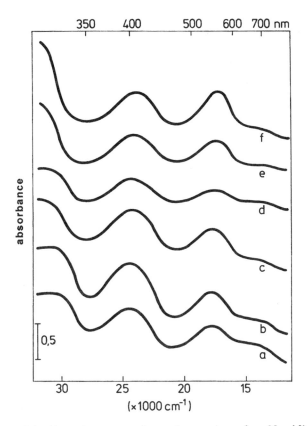

Figure 7.4 Absorption spectra of corundum specimens from Nepal [10].

coloured specimens amounting to as much as 4% [1], but there is a certain dependence on the particular lattice doped with these ions. It is also connected with the principle of electrical neutrality. If during the synthesis of rubies MgO is introduced, Mg^{2+} ions, which replace Al^{3+} ions, cause a deficit of one positive charge (per Al^{3+} ion). As a result, there is a tendency to achieve neutrality, which is possible if, for instance, Cr(III) is oxidised to Cr(IV), i.e. corresponding to each $2Al^{3+}$ pair there is a $Cr^{4+} + Mg^{2+}$ system. This has an effect on the colour, namely that instead of the usual red ruby, one obtains ruby which is brown orange, owing precisely to the presence of Cr(IV). A mineral of this kind occurs naturally as padparadschah corundum. Let us recall here that CrO_2 is dark brown, which is also the colour of other Cr(IV) compounds, including the complex compounds synthesised and studied by one of the present authors [11]. The occurrence of Cr^{4+} ions has recently been found in garnet-type gadolinium–scandium–gallium crystals and aluminium–yttrium crystals with Cr^{3+} and Mg^{2+} impurities [12].

The question of the spectroscopic properties of Cr(III) is not limited to corundums (ruby and sapphires), as this element also occurs in other minerals, which show very well formed spectra of this ion. One such very precious gemstone is emerald, whose

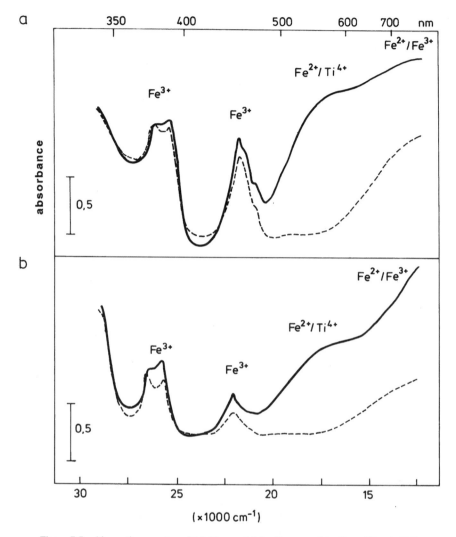

Figure 7.5   Absorption spectra of (a) blue and (b) yellow sapphire from Nigeria [10].

colour has been discussed above. Figure 7.7 [4] shows the polarised spectrum of this mineral (from Columbia).

The dichroism of transitions results from the deformation of a cubic structure and the general character of the spectrum is typical of the hexacoordinate structure of Cr(III) compounds. The two main absorption bands have, of course, slightly different positions depending on σ and π polarisation. As can be seen from the chromaticity diagram, all minerals discussed here are green (Figure 7.8) [4].

In addition to Cr^{3+}, the minerals discussed in this section contain Fe^{2+}, V^{3+}, Ti^{3+}, and Mn^{3+} ions, and their particular amount in each case may cause the occurrence of various shades of colour – and different chromaticity coordinates.

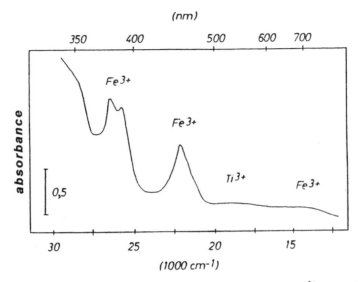

Figure 7.6 Absorption spectrum of a sapphire specimen without Cr^{3+} ions [10].

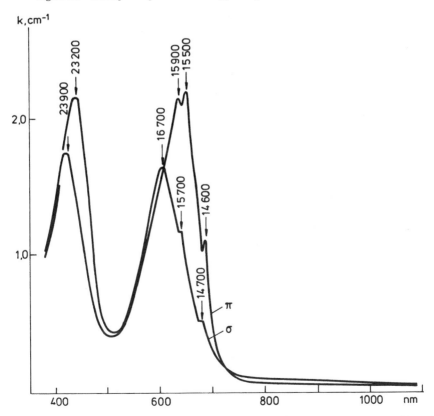

Figure 7.7 Polarized absorption spectrum of emerald [4]; k (in cm^{-1}) denotes absorption coefficient, here and in many subsequent Figures.

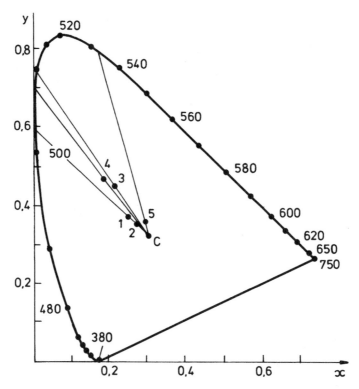

Figure 7.8 CIE chromaticity diagram of emeralds from various sources (see ref. [4] for details).

7.2.2 Electronic spectra and colour related to the occurrence of iron ions

The situation connected with the occurrence of iron ions as impurities is exceedingly complicated. The ions may occur as Fe^{2+} or Fe^{3+}, as Fe^{2+}–Fe^{3+} pairs, in octahedral or tetrahedral environments, or even with coordination number 8, and finally as pairs with other ions, e.g. with Ti^{4+} (as mentioned earlier).

Some spectra of beryl with a colouration due to the presence of iron are presented in Figure 7.9 [4].

Nine different transitions have been identified in the optical spectra of iron-bearing beryls. Of those nine, essential from the point of view of colour are the absorption bands in the visible range at about $24\,300\,\text{cm}^{-1}$ (ca. 471 nm), at approximately $12\,400\,\text{cm}^{-1}$ (ca. 806 nm), at 16 400 (ca. 608 nm) as well as a series of low-intensity bands in the visible range and the near ultraviolet. This means that depending on the particular positions and intensity relations among the bands, there may be very different shades of colour.

Iron occurs as an allochromatic element in, for instance, olivine-type orthosilicates with a general composition of the form $(Mg, Fe)_2SiO_4$. The first member of this group of minerals is forsterite, $Mg_2[SiO_4]$, and the last is fayalite, $Fe_2[SiO_4]$. The spectra of these minerals are presented in Figure 7.10 a, b [2].

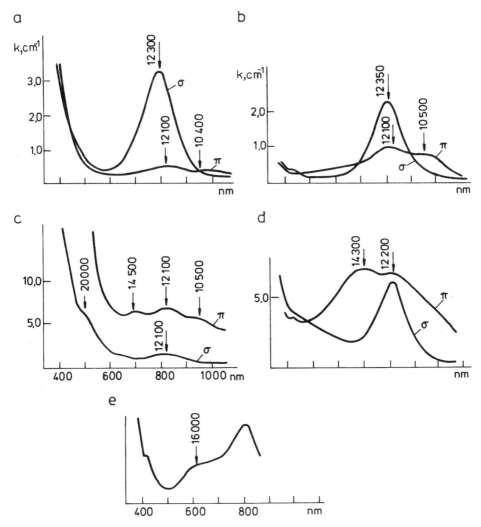

Figure 7.9 Absorption spectra of iron-containing beryl specimens from various sources, for σ and π polarization [4].

Owing to the crystallographic structure, there are three components (three different spectra), α, β, and γ, in the polarised spectra. Greatest transmittance can be observed in the 550–700 nm range for all spectra, both those of forsterite and those of fayalite. The absorption bands are well shaped and undergo a bathochromic shift as the concentration of iron ions increases. According to Burns [2], the shifts, expressed in wavenumbers, are linearly dependent on the Mg^{2+}–Fe^{2+} composition. In the crystallographic structure of these minerals, there are two deformed octahedral centres denoted as M_1 and M_2, with local symmetries (approximately) D_{4h} for M_1 and C_{3v} for M_2. Bands at about 1000 nm are ascribed to Fe^{2+} in M_2, while bands at approximately 850–900 nm are ascribed to the same ions in M_1.

One of the principal types of minerals whose absorption spectrum can conveniently be explained by the presence of Fe^{2+} ions are garnets of the almandine–pyrope series. Figure 7.11 [2] presents the spectrum of almandine containing 70% of $Fe_3Al_2(SiO_4)_3$. Strong light transmittance occurs in the 700–900 nm range, while the characteristic absorption bands have wavenumbers corresponding to near infrared, the furthest at $4450\,cm^{-1}$ (about $2.23\,\mu m$). Strong transmittance in the indicated range causes the occurrence of red, orange, and pink colours in many garnets. The chromaticity diagram from publication [4] (Figure 7.12), comprising 35 minerals, shows where the colours are situated.

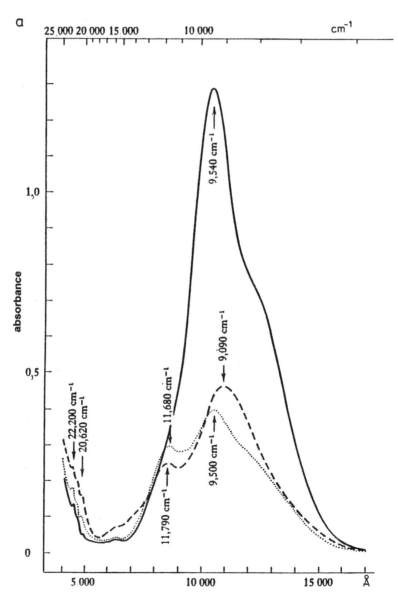

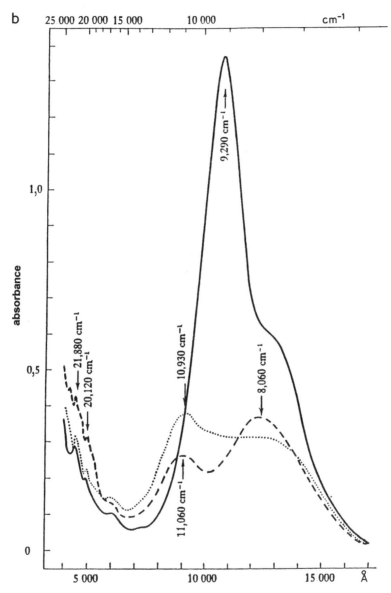

Figure 7.10 Polarized absorption spectra of olivines: .... α spectra; - - - - β spectra; —— γ spectra. (a) Forsterite, from Jan Mayan ($t = 0.5$ mm); (b) fayalite, from Rockport, Mass. ($t = 0.08$ mm). Optical orientation: $\alpha = $ b; $\beta = $ c; $\gamma = $ a. From ref. [2].

The colour of almandine gemstones, normally deep red or brownish black, may also be orange red, pink red, red, or red purple. Marked variability of the iron(II) content is one of the main determining features, though according to Płatonow [4] increasing absorption in that short section of the visible range which causes a red or orange appearance is also due to an $O^{2-} \rightarrow Fe^{3+}$ charge-transfer transition.

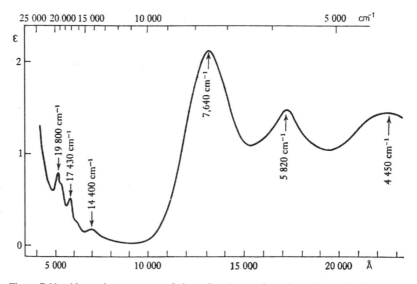

Figure 7.11   Absorption spectrum of almandine (garnet from Fort Wrangell, Alaska) [2].

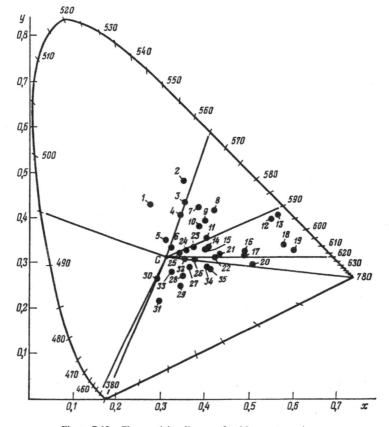

Figure 7.12   Chromaticity diagram for 35 garnet specimens.

7.2.3   Electronic spectra and colour of minerals containing manganese

The approximate number of manganese-containing minerals is 150. Minerals that owe their colour mostly to Mn^{2+} ions include rhodonite, pyroxmangite, (Fe, Mn)$_7$[Cr$_7$O$_{21}$], bustamite, (Mn,Ca)$_3$[Si$_3$O$_9$], johannsenite, (Ca,Mn)[Si$_2$O$_4$], and spessartine; those that owe their colour to Mn^{3+} include epidote, andradite, grossular, and idocrase. Other manganese-containing minerals are pyrolusite, rhodochrosite, and manganite.

One of the minerals whose colour is mainly due to the presence of manganese ions is kunzite. This belongs to the spodumene group and is described by the general formula LiAlSi$_2$O$_6$. Interpretation of electronic spectra and an unequivocal explanation of the cause of the colour of this mineral are very complicated, though they are aided by additional studies including luminescence, EPR spectra, and the effect of X-radiation on the occurrence of shades of colours.

The important factors are oxidation states, coordination numbers, and symmetry. Three oxidation states should be considered, namely +2, +3, and +4, and both octahedral and tetrahedral configurations. The spectroscopy of Mn^{2+}, Mn^{3+}, and Mn^{4+} is very well established both for the solid phase and for solutions [7, 13].

It should be borne in mind that high-spin Mn^{2+} in an octahedral ligand field has the $3d$ electronic configuration $t_{2g}^3 e_g^2$, so that d-d electronic transitions are doubly forbidden (Laporte- and spin-forbidden). Molar extinction coefficients (ε) are thus an order of magnitude smaller, and colours correspondingly even weaker, than for other d^q configurations. Typically ε values are in the region of 0.1 to 0.2 $M^{-1} \cdot cm^{-1}$. The value of $10Dq$ for the Mn^{2+} hexa-aqua ion is 8000 cm^{-1}. Aqueous solutions containing high concentrations of manganese(II) salts are pink, as are e.g. freshly precipitated hydrated manganese(II) sulphide, MnS, familiar from qualitative analysis, and manganese(II) fluoride, MnF$_2$. A typical absorption spectrum of manganese(II), as MnF$_2$, is shown in Figure 7.13 [6]; the spectrum of an aqueous

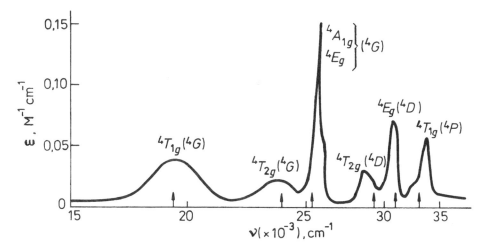

Figure 7.13   Absorption spectrum of MnF$_2$ [6].

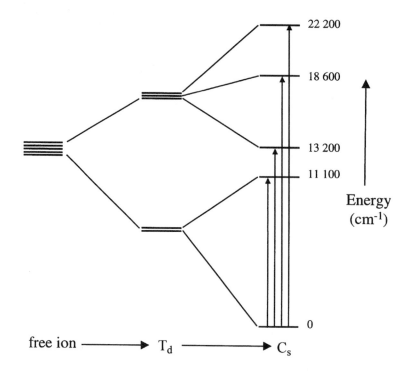

Figure 7.14   The *d-d* transitions for Mn^{3+} centres in pink clinozoisite [15].

solution of manganese(II) perchlorate is almost identical [14]. All the absorption bands appearing in such spectra are due to sextet-quartet transitions.

In the case of Mn^{3+} ions, with their d^4 configuration, we expect one allowed quintet-quintet transition, $^5E_g \rightarrow {}^5T_{2g}$, in O_h symmetry. In the aqua-complex of this symmetry the wavenumber of the relevant band is $21\,000\,cm^{-1}$ (476 nm). However in many solids the environment of the Mn^{3+} is much less symmetrical, with tetragonal splitting (rhombic symmetry) giving up to four transitions. Thus, for example, four bands assignable to *d-d* transitions of the Mn^{3+} in distorted tetrahedral environments have been observed in piemontite $\{Ca_2Al_{(3-x-y)}Mn_xFe_y(OH)(Si_3O_{12})\}$, whose colour varies from lightish red through reddish-brown to black, and in pink Bulgarian clinozoisites {of almost identical composition, viz. $Ca_2Al_{3-x}Mn_x(OH)(Si_3O_{12})$} (Figure 7.14) [15].

For the d^3 electronic configuration of the Mn(IV) ion, electronic spectra may be similar to those of Cr(III), but thanks to its greater formal charge, $10Dq$ for the Mn(IV) aqua-ion is approximately $28000\,cm^{-1}$. We can thus expect a substantial hypsochromic shift of absorption bands in comparison with the spectra of Cr(III) compounds.

In the case of natural kunzite it has been found [4] that three different polarised spectra can be obtained; Figure 7.15 gives the polarised spectra of the natural green mineral, while Figure 7.16 presents the spectra of pink kunzite.

Both zones of the natural mineral contain manganese ions and the spectra are in total conformity with the spectra of separate green and pink minerals. The authors identified three main absorption bands at $11\,400$, $16\,000$, and $18\,700\,\mathrm{cm}^{-1}$, the former two characterising the green variety and the last characterising the pink variety. The intensity of the colour is different for different polarised spectra, being the greatest for γ polarisation both in the case of green and pink colours. On the

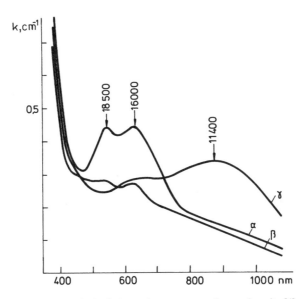

Figure 7.15   Polarized absorption spectrum of green kunzite [4].

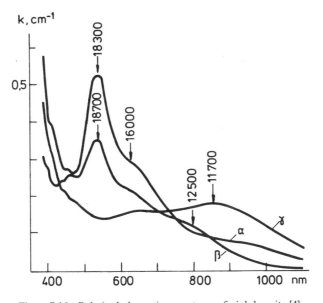

Figure 7.16   Polarized absorption spectrum of pink kunzite [4].

basis of other experimental data, and particularly changes which occur as a result of X-radiation and in chemical examinations, the authors of the monograph cited above conclude that the pink colour is related to the presence of Mn^{3+} ions in a particular position in the crystal, where they replace Al^{3+} ions in the position designated as $M1$. The same situation also occurs in other minerals, for instance in tourmaline. It is assumed that the symmetry is lowered to C_{2v} and there are four bands, one of which may have its wavenumber below $2000\,cm^{-1}$.

The absorption bands of the green variety, at 11 400 and $16\,000\,cm^{-1}$ (625 nm), correspond to identical bands for the Mn^{3+} ion, but in the position described as $M2$ in the structure of spodumene. As transition energies are lower than for the pink variety, the crystal field strength is naturally weaker, and Mn^{3+} ions take the positions of Li^{+} ions.

The mechanism of colour formation in spodumenes according to Płatonow et al. [4] is as follows:

$$
\begin{array}{ll}
\text{colourless centres} & Mn^{2+}(M1) + Mn^{2+}(M2) \\
\text{green centres} & Mn^{2+}(M1) + Mn^{3+}(M2) \\
\text{pink centres} & Mn^{3+}(M1) + Mn^{2+}(M2)
\end{array}
$$

7.3 COLOUR AND ELECTRONIC SPECTRA OF SOME MINERAL PIGMENTS

Mineral pigments belong to the important, widely used, and long-familiar group of coloured materials known as pigments, dyes, and stains. In general the term pigment is applied to insoluble inorganic substances, whereas dyes are generally soluble organic compounds.[1] Some pigments occur in nature as minerals. Examples include crocoite, $PbCrO_4$, and uvarovite, Cr_2O_3. Others consists of simple synthetic inorganic compounds, and more complex species such as metal phthalocyanines.

Mineral pigments have been in use for an extremely long time. The paintings in the caves at Lascaux incorporate hematite, goethite, and lepidocrocite, α-Fe_2O_3, α-$FeOOH$ and γ-$FeOOH$ respectively, for red, yellow, and orange colouring, and hausmannite (Mn_3O_4) as the black pigment. The ancient Greeks used Fe_3O_4, which they prepared from Fe_2O_3, as a black pigment in decorating vases. The ancient Egyptians also used oxides of iron as pigments (and, in one of the earliest uses of inorganic pharmacology, to increase virility and potency), and developed several blue pigments. These included the so-called Egyptian blue, whose colour stems from the incorporation of Cu^{2+} into a calcium silicate lattice, and lapis lazuli. Azurite blue was also in use around this time, but was found to suffer from the disadvantage of turning green under certain conditions. These and later blue pigments are listed in Table 7.3.

Mineral pigments are usually classified according to their colour and the colour they impart to such commercial products as paints, varnishes, inks, and synthetic

---

[1] Ruthenium red, $[(H_3N)_5Ru-O-Ru(NH_3)_4-O-Ru(NH_3)_5]^{6+}$, is a (rare) example of an inorganic dye; OsO_4 is a stain much used in biology – it is reduced in use to give the black pigment osmium metal.

Table 7.3 A chronology of blue pigments

BC	~2000	Egyptian blue	$(Cu,Ca)Si_4O_{10}$
		Ultramarine	ca. $CaNa_7Al_6Si_6O_{24}SO_4$
		Lapis lazuli[a]	
		Azurite	$Cu_3(CO_3)_2(OH)_2$
AD	~0	Maya blue	Indigo-doped palygorskite
	1704	Prussian (Berlin) blue	$Fe_4[Fe(CN)_6]_3.nH_2O$
	1903	Manganese blue	$BaMn^VO_3(OH)$ (+ $BaSO_4$)
	1948	Vanadium blue	V^{IV} doped into SiO_4 sites

[a] The colour of lapis lazuli is due to lazurite, haüynite, and sodalite. The lapis lazuli in some ancient artefacts has been patched up with (the much cheaper) indigo.

chemical materials. There are pigments of the colours yellow, orange, brown, red, blue, violet, and green – almost corresponding to the colours of the rainbow. Black, grey, and white are also recognised as pigments. Table 7.4 gives examples of mineral pigments, arranged according to their colour; a small selection of pigments are illustrated in colour in Figure C11. As is apparent from Table 7.4, some pigments are single compounds, but many are mixtures of two or more ingredients. Pigments may be mixed in order to obtain exactly the required colour, or a coloured pigment may be mixed with a white pigment to increase opacity – for pigments are employed both for obtaining a specific colour and to render the surface of an object non-translucent.

The technical requirements are different for these two aspects. For non-translucency the desired qualities are good diffusion and reflection of light from the surface [6]. The situation in regard to colour is more complicated [1], though it is well-established that the best pigments are characterised by a high refractive index, for example, chrome green with a value of 2.5. Further examples of pigments with high refractive indices are presented in Table 7.5.

An essential technical requirement for pigments used to colour ceramics is that they withstand the high temperatures, of the order of 1000 °C, of the firing process. Recent developments involving new materials with the perovskite ($CaTiO_3$) structure, such as $SrSnO_3$ or $CaZrO_3$, have yielded pigments stable to temperatures as high as 1200 or even 1300 °C.

The occurrence of colour in mineral pigments involves processes detailed earlier, including ligand field transitions, as in Cr_2O_3, MLCT transitions, as in $PbCrO_4$ intervalence transitions, as in $Fe_4[Fe(CN_6)]_3$, and interligand transitions, as in ultramarine (see Section 7.3.4 below). Electronic (reflectance) spectra of some red, green, and blue pigments in lithium carbonate matrices are shown in Figure 7.17.

Inorganic substances used as luminophores in television screens are in a sense related colour sources. Mixtures used for this purpose include $ZnS + Ag$ (blue), $ZnCdS_2 + Ag$ (green), and $Y_2O_3 + Eu$ (red). This topic is also briefly alluded to in Chapter 4 (see Section 4.3).

We shall now discuss a selection of coloured mineral pigments, arranged according to their colours [16]. Reference [17] lists 51 inorganic pigments; fuller details of the production, properties, and uses of inorganic pigments may be found in references [18] and [19]. The latter review covers 127 pages, giving 281 references.

Table 7.4 Some metal-containing pigments[a]

Colour	Pigment	Formula
Yellow[b], orange	Iron oxide yellow	FeO(OH)
	Cadmium yellow	CdS[c]
	Cadmium orange	$CdS_{1-x}Se_x^d$
	Zinc yellow	$ZnCrO_4$
	Chrome yellow	$PbCrO_4 + (PbSO_4)$
	Chrome orange	$PbCrO_4 + PbO$
	Molybdenum orange	$PbCrO_4 + PbMoO_4 + (PbSO_4)$
Brown, red	Hematite, limonite, siderite	$Fe_2O_3(nH_2O)$
	Sienna	Iron + manganese oxides
Red	Cadmium red	$CdSe_xS_{1-x}^d$
	Red lead	Pb_3O_4
	Vermilion	HgS
Blue, violet, purple	Azurite	$Cu_3(CO_3)_2(OH)_2$
	(Thénard's) cobalt blue	$CoAl_2O_4$
	Prussian (Turnbull's) blue[e]	$Fe_4[Fe(CN)_6]_3$
	Copper (blue) phthalocyanine[f]	
	Lazurite	$(Na,Ca)_8[(SO_4,S,Cl)_2(AlSiO_4)_6]^g$
	Zircon blue[h]	$(Zr,V)SiO_4$
	Purple of Cassius	tin-stabilised colloidal gold
Green	Chrome oxide green[i]	Cr_2O_3
	Malachite	$Cu_2CO_3(OH)_2$
	Scheele's green	$CuHAsO_3$ or $CuAs_2O_4$
White	Titanium dioxide	TiO_2
	Zinc oxide	ZnO
	Lead white[j]	$Pb_3(CO_3)_2(OH)_2$
Grey	Zinc primer	Zn in silicate matrix
Black	Iron oxide black	Fe_3O_4
	Osmium[k]	Os
	Carbon	C

[a] A range of colours, including blue, yellow, pink, etc. can be obtained by doping zirconium silicate with various transition metal ions. [b] The bright yellow pigment obtained by doping a zirconium-based lattice with praseodymium oxide has already been mentioned, in Section 4.3. [c] $Cd_{1-x}Zn_xS$ gives primrose pigments. [d] Yellow → orange → red as x goes from 0 to 1; an alternative method of varying colour yellow → orange → red is to add increasing amounts of HgS to CdS. [e] Identical to Turnbull's Blue (see p. 187); various iron blues are all closely related hexacyanoferrates. [f] The so-called β-form is green; a range of blue → green pigments can be generated by the use of various metal phthalocyanines and related compounds. [g] May also contain chloride. [h] V^{4+} doped into $ZrSiO_4$ or into $ZrGeO_4$ gives blue and purple respectively. [i] Also known as Gingnet's Green, and as Chrome Green – the latter name also being used for pigments consisting of mixtures of chrome yellow and iron blue (footnote e) (cf. Section 7.3.3). [j] Other white pigments, of slight but significantly different appearance, include $BaSO_4$ and lead aluminate (but TiO_2 and ZnO are by far the most important white pigments). [k] Osmium may be considered a pigment in consequence of its biological use.

7.3.1 Red and brown pigments

The most important members of this group are compounds of iron, especially the various oxides of this element. Iron oxide pigments account for about 40% of the

Table 7.5 Some mineral pigments which have a high index of light refraction

Pigment	Formula	Refractive index	Colour
Vermilion	HgS	3.0	red
Red lead	Pb_3O_4	2.5	red
Chrome green	Cr_2O_3	2.5	green
Cadmium yellow	CdS	2.4	yellow

total production of coloured inorganic pigments. Depending on the precise method of preparation and composition, these pigments may show various hues, viz. light yellow, dark yellow, red, brown, and chestnut. The principle natural sources of these pigments are hematite, limonite, magnetite, pyrite, and siderite. Hematite, limonite, and siderite are usually used to obtain red pigments; commercial products contain varying amounts of Fe_2O_3, up to about 90%. Such pigments also contain SiO_2,

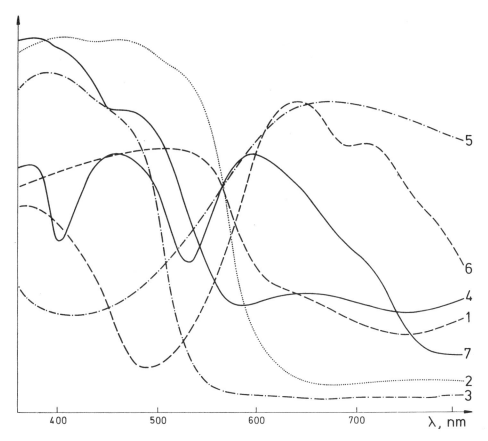

Figure 7.17 Reflectance spectra of some mineral pigments in Li_2CO_3 matrices: 1 – red iron; 2 – molybdenum orange; 3 – chrome yellow; 4 – iron yellow; 5 – iron blue; 6 – copper phthalocyanine; 7 – chrome (oxide) green.

Al_2O_3, CaO, and MgO. The colour of Fe_2O_3 is formally due to ligand field transitions in the Fe^{3+} (Table 7.3). However, for the d^5 high-spin electronic configuration there are no spin-allowed transitions, though spin-forbidden transitions are in fact observed in the visible region. The red colour is probably partly due to an LMCT transition, from O^{2-} to Fe^{3+}. The reflectance spectra of red pigments (see Figure 7.17) show strong reflection in the 620 to 700 nm range, and very weak reflection in the 380 to 500 nm range.

Synthetic iron oxides also play an important role in the production of red pigments, as do red lead, Pb_3O_4, and cadmium red, $CdSe_{1-x}S_x$. The colour of the last-named changes, from light red to red to dark red, as the proportion of selenium increases.

A recently described brown pigment is derived from vanadium blue – it contains V^{4+} doped into Zn_2GeO_4 (see Section 7.3.4 below).

7.3.2   Yellow and orange pigments

The main compounds for the production of these colours are hydrated iron oxides, lead chromate and molybdates, and zinc and molybdenum yellows.

The principal natural mineral pigments in this group include ochre and sienna, obtained from limonite ores. Their composition ranges from $Fe_2O_3.2H_2O$ to $Fe_2O_3.4H_2O$, with SiO_2 and Al_2O_3 occurring as impurities. They may be light and bright yellow, green or brown red, or even black brown. The approximate composition of synthetic iron pigments in this group is $Fe_2O_3.H_2O$.

The reflectance spectra of 3 mineral pigments, chrome yellow, iron yellow, and molybdenum orange are shown in Figure 7.17.

Yellow and orange pigments show quite similar reflectance spectra. In the former, more than 80% reflection occurs in the 540–700 nm range; in orange pigments this range is narrower, 600–700 nm. At the same time, reflection in the blue green region is markedly stronger in yellow pigments than in orange ones.

Chrome yellows, based principally on lead chromate, are among the most important synthetic coloured pigments. They show hues from light yellow green to reddish yellow. In this group, there are also low-saturation pigments, e.g. lemon-coloured, as well as medium-saturation yellows. The latter are practically pure $PbCrO_4$. Generally, chrome yellows are 'pure' with a strong lustre.

Chrome oranges are pigments showing hues from pale red to deep orange red. The foundation of their composition is basic lead chromate with the formula $PbO.PbCrO_4$.

7.3.3   Green pigments

Green-coloured pigments are based virtually exclusively on chromium compounds. Three subgroups may be distinguished: so-called chrome green, pigments based on Cr_2O_3 (chrome oxide green), and those based on hydrated chromium(III) oxide.

Chrome greens are typical mixed mineral pigments, where the green colour is obtained by suitably mixing two pigments (compounds), namely chrome yellow, $PbCrO_4$, and iron blue, with the composition $[Fe(NH_4)][Fe(CN)_6]$. Depending on the proportions of the two components, pigments of different coloration are obtained (i.e. both pigments with various chromaticity coordinates, x and y, and pigments of different luminance). Very light shades correspond to pigments containing about 2% of iron blue; the darkest shades may hold as much as approximately 64% [16].

The obtaining of mixed pigments with definite chromaticity parameters is a complex issue, discussed in more detail in numerous monographs and articles (e.g. [20]). Generally it can be stated that in this case colorimetric properties are the result of subtractive colour mixing. This means that from the spectroscopic point of view, in the electronic spectrum of a mixed pigment, there are regions (maxima) of absorption and transmission like in the individual ingredients (of course in some cases those characteristic regions may be disturbed or distorted). Chemically, the mixing of pigments should not lead to chemical reactions, which could result in a . product with unpredictable and undesirable colour properties. Pigments based on Cr_2O_3 generally show a narrow range of hues, from yellowish green to dark blue green. They are usually less lustrous.

The reflectance spectrum contour of chrome oxide, Cr_2O_3, is the most typical of Cr(III) compounds. As is well known, because there is no colour complementary to green in the spectrum, the occurrence of green colour is due to the additive mixing of blue and red or yellow. This means, in other words, that in the spectrum of a green substance, there are at least two absorption bands and at least one high transmission or reflectance region. This can be seen in Figure 7.17.

7.3.4 Blue, violet, and purple pigments

As can be seen in Table 7.3, there are many known pigments with this range of colours. In the case of blue pigments two main groups may be distinguished, one based on iron compounds, the other on ultramarine. Several blue pigments have been known for a long time, indeed ultramarine (lapis lazuli) was known to the ancient Egyptians (Table 7.3). Another long-established pigment is Prussian Blue (Berlin Blue), which is hydrated iron hexacyanoferrate, $Fe_4[Fe(CN)_6]_3.16H_2O$. Turnbull's Blue, which is prepared by adding Fe^{3+}aq to $[Fe^{II}(CN)_6]^{4-}$, is identical with Prussian Blue, prepared by adding Fe^{2+}aq to $[Fe^{III}(CN)_6]^{3-}$. Although this pigment is a delocalised (Robin and Day Class II, see Section 3.3.3) mixed-valence compound, it may be formulated as ferrous ferricyanide, $Fe^{II}_4[Fe^{III}(CN)_6]_3.16H_2O$. In the crystal structure of the solid phase, Fe^{II} and Fe^{III} ions are both in octahedral symmetry, but are in different ligand environments. The Fe^{II} ions are bonded to the carbon atoms, while the Fe^{III} ions are partly bonded to nitrogen atoms of cyanide ligands and partly bonded to oxygen atoms of water molecules, of which there are sixteen per mole of compound. In other words the Fe^{III} ions are in an environment of lower symmetry than the Fe^{II} ions. In this situation charge-transfer may take place:

$$Fe^{2+} + {}^*Fe^{3+} \longrightarrow Fe^{3+} + {}^*Fe^{2+}$$

It is this charge-transfer transition which imparts the intense blue colour to this compound, which is very important from the point of view of its use as a pigment in a number of industrial applications, especially in the production of inks. The reflectance of this blue iron pigment is shown in Figure 7.17. The spectra of all pigments in this group have very similar contours, whose chief characteristic is high reflectance in the range 420 to 480 nm. The topic of electronic transitions in metal cyanide complexes has been discussed many times (see, e.g. [1, 5, 6, 13]).

Ultramarine shows a characteristic reflectance spectrum similar to that of Prussian blue. However the situation here is more complicated as ultramarine is a very complex substance, of approximate composition $Na_7Al_6CaSi_6O_{24}SO_4$. It may be regarded as derived from sodalite, $Na_4Al_3Si_3O_{12}Cl$, with the key difference that sulphur anion radicals replace the chloride of sodalite. Synthetic ultramarine may be blue, green or even red, depending on the nature of these sulphur radicals. Their characteristics were probed by Raman spectroscopy many years ago [21], and discussed in detail by Płatonow *et al.* some 15 years ago [4]. More recently Reinen has given an interesting and valuable account of the nature of chalcogen colour centres in ultramarine-type solids [22]. The colours of ultramarines depend on the relative concentrations of the $S_2^{-\bullet}$, $S_3^{-\bullet}$, and $S_4^{-\bullet}$ anion radicals. $S_3^{-\bullet}$ imparts the blue colour (band at 600 nm) characteristic of the original pigment ultramarine and the mineral lapis lazuli. $S_2^{-\bullet}$ imparts a yellow colour (band at 400 nm); $S_2^{-\bullet}$ and $S_3^{-\bullet}$ together give green colours, while $S_4^{-\bullet}$ is responsible for the colour of red ultramarines. Figure 7.18 presents powder reflection spectra of two sulphur-containing ultramarines with differing $S_2^{-\bullet} : S_3^{-\bullet}$ ratios, one violet blue, the other steel blue. ESR spectroscopy has provided evidence for these radical anions being very tightly bound into the lattice, which helps to explain the high colour stability of ultramarine [23]. In recent years new methods for the preparation of ultramarines, generally more environmentally-friendly than the classic 19th century industrial processes, have been developed.

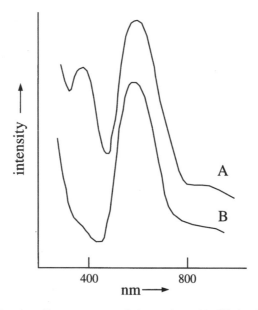

Figure 7.18   Powder reflectance spectra of ultramarines with differing $S_2^{-\bullet} : S_3^{-\bullet}$ ratios.

These give a greater range of shades and colours. It is also possible to incorporate $Se_2^{-\bullet}$ radical anions, giving red ultramarine, or $Te_2^{-\bullet}$, giving green or blue colours.

Lazurite is the naturally occurring mineral which corresponds to synthetic ultramarine. According to Płatonow *et al.* [4] the colour of lazurite is due to the paramagnetic centres $S_2^{-\bullet}$, $S_3^{-\bullet}$, and $SO_4^{-\bullet}$; more recently the colour of lazurite has been assigned just to the $S_3^{-\bullet}$ and $SO_4^{-\bullet}$. Their stability, and hence the stability of the blue colour, is attributed to a hole mechanism consequent on electron loss.

The group of blue pigments also contains copper(II) phthalocyanine, which exhibits very intense absorption. A solution in 1-chloro-naphthalene has its main absorption band at 678 nm, with $\varepsilon = 21\,800\,M^{-1}{\cdot}cm^{-1}$, and a band at 350 nm with $\varepsilon = 57\,540\,M^{-1}{\cdot}cm^{-1}$[24]. Copper(II) phthalocyanine exhibits strong brightness due to the small half-width of the band in the visible region, which is between 300 and $700\,cm^{-1}$ (in solution). The reflectance spectrum of this pigment has been included in Figure 7.17 above.

Vanadium blue is a relatively recent (1948) addition to the range of blue pigments [17]. It contains V^{4+} ions doped into tetrahedral positions in a silicate lattice; in zircon blue the host lattice is $ZrSiO_4$. The blue colour changes to purple if the V^{4+} ions are doped into $ZrGeO_4$ (or $HfGeO_4$). The isoelectronic (d^1) Cr^{5+} ion can in analogous manner be incorporated into tetrahedral sites in phosphate lattices to give a range of blue and green colours. The tetrahedral environments of these d^1 ions are distorted by compression or elongation; in each case the symmetry is reduced from T_d to D_{2d}. Either way the single d-d transition is split into two components, with the degree of distortion controlling the transition energies (Figure 7.19) and thus the colour. The relatively low symmetry of these D_{2d} sites of course results in less forbidden transitions than in analogous octahedral systems, and hence in more

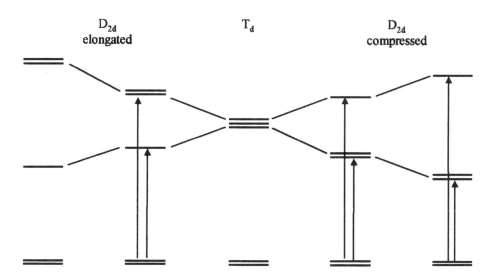

Figure 7.19 The energy level splitting consequent on $T_d \rightarrow D_{2d}$ distortion in transition metal ML_4 species.

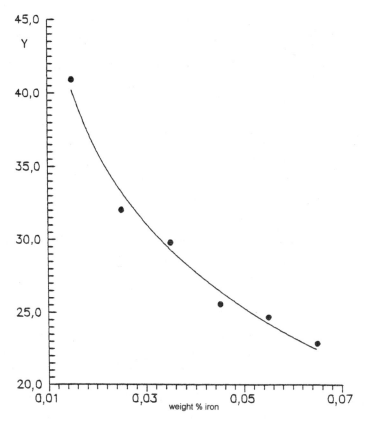

Figure 7.20 Dependence of luminance Y on weight per cent iron doped into Li_2CO_3.

Table 7.6 CIE and CIELAB chromaticity coordinates and λ_d of some pigments (applying the Kubelka-Munk method)

No.[a]	x	y	Y	L^*	a^*	b^*	λ_d (nm)
1	0.2085	0.2057	0.8204	23.411	12.117	−15.209	475.0
2	0.5931	0.3466	12.5270	58.042	54.177	45.771	607.0
3	0.4901	0.4630	69.3189	102.662	16.193	96.777	580.5
4	0.4547	0.4260	14.8288	61.399	10.443	38.543	582.0
5	0.1567	0.3551	1.5058	28.645	−27.280	−3.996	493.5
6	0.4261	0.3383	4.2921	40.615	17.225	9.691	600.0
7	0.2937	0.4064	5.6987	44.639	−16.783	9.630	535.0

[a] 1 – iron blue; 2 – molybdenum orange; 3 – chrome yellow; 4 – iron yellow; 5 – copper phthalocyanine (phthalocyanine blue); 6 – iron red; 7 – chrome oxide green.

strongly coloured pigments. V^{4+} doped into Zn_2GeO_4 gives a brown colour, as does Fe^{3+} doped into $ZnGa_2O_4$. Now the colour is due to a charge-transfer band whose cut-off, which can be varied by varying the Fe:Ge ratio, is in the visible region.

Table 7.7 The dependence of values of chromaticity coordinates on concentration of iron red

C weight ratio	x	y	Y	L^*	a^*	b^*
0.01	0.3648	0.3338	40.9631	86.150	17.699	8.984
0.02	0.3802	0.3357	32.0524	79.386	20.575	11.025
0.03	0.3826	0.3361	29.8115	77.491	20.715	11.217
0.04	0.3905	0.3362	25.5996	73.655	21.919	11.795
0.05	0.3948	0.3395	24.7137	72.795	21.826	13.078
0.06	0.3966	0.3378	22.8838	70.952	22.334	12.578

7.3.5 Chromaticity of mineral pigments

Evaluation of the properties of coloured pigments is based both on examination and analysis of their electronic spectra – absorption, transmission, or reflectance – and on the determination of their chromaticity parameters. The examination of the electronic spectra provides information not only on wavelength (wavenumber) ranges for absorption or reflection, but also on intensities of absorption bands or on the percentage of light reflected. Organic dyes are characterised by absorption bands with very high molar extinction coefficients, which can be as high as $10^5\,M^{-1}{\cdot}cm^{-1}$. Extinction coefficients for mineral pigments are normally considerably lower, generally of the order of 10^3 to $10^4\,M^{-1}{\cdot}cm^{-1}$. Such values are characteristic of pigments that owe their colour to charge-transfer transitions. Investigations of new or of familiar dyes and pigments also rely on values characterising the chromaticity of such substances. These include both the x, y, and Y coordinates in the CIE system and the L^*, a^*, and b^* in the CIELAB system. Also of practical importance are the set of dominant (or complementary) wavelengths, the purity of excitation, and, in addition to luminance, the hue angle h_{ab}. For the purpose of comparing several pigments, colour differences expressed as ΔE_{ab}^* or, more often, ΔH_{ab}, are employed.

Some experimental results obtained in studies of seven mineral pigments used in the paint and varnish industries are presented in Figures 7.17 and 7.20 and in Tables 7.6 and 7.7 [25]. The reflectance spectrum of the pigment chrome oxide green, mentioned above, which is curve 7 of Figure 7.17, shows the most characteristic contour entirely consistent with the spectrum of α-Cr_2O_3, which was given in Figure 3.35. It is worth noting that only rarely are all three spin-allowed transitions observable in the electronic spectra of Cr(III) compounds. In the present case the positions of the three bands are at approximately 580, 470, and 375 nm.

Blue pigments such as iron blue and copper phthalocyanine show strong reflectance in the 400 to 500 nm range, which determines their colour. However there are significant differences within this range, which makes it possible to distinguish between them through differences in hue, as is confirmed by the chromaticity diagram.

A similar situation is observed for two yellow pigments, ferrite yellow and chrome yellow, and even for molybdenum orange. They all show strong reflectance in the 600–700 nm range, especially molybdenum orange and chrome yellow.

The chromatic properties represented in the form of the now familiar quantities are presented in the tables above.

Table 7.6 gives the CIE and CIELAB chromaticity coordinates and the dominant wavelength for the examined pigments (for a particular mass fraction) in the Li_2CO_3 matrix. These data provide a good characterisation of chromatic properties, and hence particular aspects of application.

As far as the blue pigments discussed above are concerned, the dominant wavelength for iron blue, 480.5 nm, corresponds to the blue–greenish blue colour range, while the corresponding value for copper phthalocyanine is 492 nm, which points to the blue green–bluish green region.

Hue differences are even more easily seen on the basis of the given a^* and b^* values. Positive values of both place the colour in the yellow–red area, while negative values of both situate the colour in the blue–green section. For $a^* > 0$ and $b^* < 0$, it is the red blue area (i.e. the purple area), whereas for $a^* < 0$ and $b^* > 0$, it is the yellow green region.

On the basis of these values, it can be concluded that copper(II) phthalocyanine in the Li_2CO_3 matrix is characterised by a stronger presence of green in the overall blue colour than in the case of iron blue. As for yellow pigments, the b^* value for chrome yellow is the greatest of all the pigments investigated and thus its yellow colour is markedly stronger than that of iron yellow, where an admixture of red is also detectable. At the same time, the dominant wavelength for these pigments is almost the same at, respectively, 577 and 579 nm.

Interesting data are also available on the effect of concentration on the chromaticity of pigment + matrix systems. Figure 7.20 and Table 7.7 indicate that as the mole fraction of red iron pigment in admixture with lithium carbonate increases, luminance decreases – but not linearly. As in the case of solutions, an increase in the concentration of the coloured substance correlates with an increase of light absorption and a decrease of transmittance or reflectance.

REFERENCES

[1]   Nassau, K., 1983, *The Physics and Chemistry of Color*, Wiley, New York.
[2]   Burns, R.G., 1970, *Mineralogical Aspects of Crystal Field Theory*, Cambridge University Press; 1993, 2nd edn.
[3]   Marfunin, A.S., 1975, *Spektroskopija, luminescencija i radiacionnyje centry w mineralach*, Idz. Nedra, Moscow.
[4]   Płatonow, A.N., Taran, M.N. and Balickij, W.S., 1984, *Priroda okraski samocwetow*, Idz. Nedra, Moscow.
[5]   Figgis, B.N., 1966, *Introduction to Ligand Fields*, Wiley-Interscience, New York.
[6]   Lever, A.B.P., 1984, *Inorganic Electronic Spectroscopy*, 2nd edn., Elsevier, Amsterdam.
[7]   Ballhausen, C.J., 1962, *Introduction to Ligand Field Theory*, McGraw-Hill, New York.
[8]   *Spectroskopija kristallov*, Izd. Nauka, Moscow, 1966.
[9]   Reinen, D., 1969, *Struct. Bonding*, **6**, 39.

[10] Kiefert, L. and Schmetzer, K., 1987, *Z. Deutsch. Gemmol. Ges.*, **35**, 113 (1986); **36**, 61.

[11] Bartecki, A., 1965, *Roczniki Chem.*, **39**, 167.

[12] Gamrasz, W.M., Żitniuk, W.A., Ochrimczuk, A.G. and Szestakow, A.W., 1990, *Nieorg. Mater.*, **26**, 1700.

[13] Jørgensen, C.K., 1962, *Absorption Spectra and Chemical Bonding in Complexes*, Pergamon, Oxford.

[14] Bartecki, A., 1987, *Chemia pierwiastków przejsciowych*, WNT, Warsaw.

[15] Taran, M.N., Platonov, A.N., Petrusenko, S.I., Khomenko, V.M. and Belichenko, V.P., 1984, *Geochem. Mineral. Petrol.*, **19**, 43.

[16] Gloger, W.A., 1963, in *Kirk-Othmer's Encyclopaedia of Chemical Technology*, 2nd edn., Wiley, New York, p. 495.

[17] Eppler, R.A., 1993, *Colorants for Ceramics*, in *Kirk-Othmer's Encyclopaedia of Chemical Technology*, 4th edn., Wiley, New York, Vol. 6, p. 877.

[18] Novotny, M., Solc, Z. and Trojan, M., 1996, *Pigments, Inorganic*, in *Kirk-Othmer's Encyclopaedia of Chemical Technology*, 4th edn., Wiley, New York, Vol. 19, p. 1.

[19] *Pigments, Inorganic*, in *Ullmann's Encyclopaedia of Industrial Chemistry*, 5th edn., VCH, Weinheim, 1992, Vol. A20, p. 243.

[20] Johnston, R.M., 1973, *Color Theory*, in *Pigments Handbook, Vol. III*, ed. Patton, T.C., Wiley, New York, p. 224.

[21] Clark, R.J.H. and Cobbold, D.G., 1978, *Inorg. Chem.*, **17**, 3169.

[22] Reinen, D. and Lindner, G.-G., 1999, *Chem. Soc. Rev.*, **28**, 75.

[23] Kowalak, S., Miluśka, M., Więckowski, A.B. and Goslar, J., 1994, *Mol. Phys. Rep.*, **5**, 221.

[24] Gordon, P.F. and Gregory, P., 1987, *Organic Chemistry in Color*, Springer-Verlag, Berlin.

[25] Bartecki, A., Tłaczała, T., Myrczek, J. and Raczko, M., 1990, *Barwne związki metali i ich zastosowanie (Coloured metal compounds and their applications)*, Report No. 19/90, Politechnika Wrocławska, Wrocław.

8. COLOUR AND ELECTRONIC SPECTRA OF GLASS DOPED WITH TRANSITION METAL IONS

8.1 GENERAL REMARKS

The colouring and practical application of coloured glass has probably been known as long as glass itself. Generally it can be said that the most common colouring substances are transition metal oxides, while lanthanide compounds (again mostly oxides) and to a certain extent uranium compounds (mostly in the +6 oxidation state) are occasionally used.

Table 8.1 lists the compounds believed to have been used historically to produce the coloured glass for so-called 'stained glass' windows for cathedrals and churches[1], and for ornaments and artefacts. Many of these compounds are still in use today in the production of coloured glasses for a variety of purposes.

Table 8.1 Compounds used historically for the production of coloured glass[a]

Colour	Compound
Blue	Co_3O_4; Cu_2O/CuO
Violet; purple	Mn_2O_3
Green	Cr_2O_3; V_2O_3
Brown; yellow	MnO
Yellow	MnO; TiO_2/CeO_2; CdS
Orange	$CdS_{1-x}Se_x$ (x small)
Red	$CdS_{1-x}Se_x$ (x larger); Au; Cu

[a] Taken from Table 7 (p. 591) of D.C. Boyd, P.S. Danielson, and D.A. Thompson, *Glass*, in *Kirk-Othmer's Encyclopaedia of Chemical Technology*, 4th edn., Wiley, New York, Vol. 12, 1994, p. 555.

---

[1] It should, however, be realised that for several centuries glass was very often coloured by painting pigments, in an appropriate matrix, onto its surface rather than by the incorporation of transition metal oxides into the glass itself. As early as the mid-14th century a silver-staining technique had been developed to produce yellow stained glass, as in the cathedral at Wells (see p. 55 of N. Pevsner, *The Buildings of England: North Somerset and Bristol*, Penguin Books, Harmondsworth, Middx, 1958).

Transition metal oxides form non-ideal solutions in glass; however, the structure of glass, also known as the vitreous state, differs in its properties and physicochemical parameters from crystalline substances. Without going into very fine details, let us recall that the vitreous state is described in thermodynamic terms as a condensed phase without long-range order in three dimensions. On the basis of X-ray analyses, it can be concluded that the structure of glass is amorphous, but at the same time, glass is – to a certain extent – an intermediate state between liquid and crystal. Furthermore, it can even be said that glass, like liquids or crystals, shows a short-range order.

It is probably this fact that makes crystal field (ligand field) theory a viable tool to interpret electronic transitions in the ions of transition metals introduced as oxides of such metals into a glass matrix. It should, however, be assumed from the very start that the situation here will be more complex than in the crystalline state or in the liquid phase. Moreover, it turns out that the composition of glass has a noticeable effect on the absorption spectrum and thus, potentially, on the colour.

On the basis of straightforward reasoning, it could be predicted that the introduced transition metal oxides should cause the formation of a uniform coordination sphere, MO_6 or MO_4, assuming coordination numbers 6 and 4 as the most frequently occurring. However, absorption spectra clearly show that, for instance in phosphate glasses [1], the maximum of the Cu^{2+} absorption spectrum is located at 840, 850, and 860 nm for, respectively, Mg, Ca, and Ba in glasses with the general composition given by the formula $M(II)O–P_2O_5$. Thus, as in other phases, the splitting of the energy levels of transition metal ions does not depend solely on the composition of the first coordination sphere.

In the case of metal oxides, one should also take into account the possibility of charge-transfer transitions. As has been said, transition energy (of the first band) may be determined on the basis of the value of optical electronegativities from Jørgensen's well-known formula. However, in the case of O^{2-} as the ligand, this value is not constant and may vary within rather broad limits. Thus, even though optical electronegativities are known for the metal ions, as they are constant for a given oxidation state and coordination, transition energy, and consequently its effect on the colour of glass, cannot be predicted.

It follows from the above discussion that doping glass with transition metal oxides causes the occurrence of colour, which in the simplest case is due to ligand field transitions, mostly spin-allowed, but also in part to charge transfer transitions. It can also be assumed that the particular composition of glass, and especially the type of chemical bonding, may have a significant effect on the pattern of energy levels in transition metal oxides and consequently on the particular colours. Of course, as in other systems, the colour of a glass will also depend on the concentration of the introduced oxide.

8.2   COLOURS AND ELECTRONIC SPECTRA OF GLASS DOPED WITH TRANSITION METAL COMPOUNDS

Table 8.2 gives the allochromatic colours obtained by doping calcium-silicate glass with transition metal ions. A table of this sort can only provide approximate

Table 8.2 Allochromatic colours of soda-lime-silicate glasses containing $3d$ ions

Transition metal ion	Electronic configuration	Colour
Ti(III)	$3d^1$	violet purple
V(III)	$3d^2$	yellow green[a]
Cr(III)	$3d^3$	green
Mn(III)	$3d^4$	purple
Mn(II)	$3d^5$	colourless
Fe(III)	$3d^5$	pale yellow green
Fe(II)	$3d^6$	blue green
Co(II)	$3d^7$	intense violet blue[b]
Ni(II)	$3d^8$	yellow, brown
Cu(II)	$3d^9$	blue, green

[a] Vanadium gives rise to a range of colours, depending on the nature of the glass and the oxidation state of the vanadium – green in silicate glass, olive green in phosphate glass, and brown in borate glass.
[b] The colour produced by cobalt oxides depends on the material of the glass – intense blue in silicate-phosphate glass, pink in low-alkali borate glass, and green in high-alkali glass.

information, even though most of the colours are either identical with or similar to those which we know for the respective oxides or for solutions containing appropriate aqua-ions or complexes. A good example is provided by cobalt(II), which in calcium-silicate glass gives an intense violet blue colour, but in phosphate glass a pink or red colour. This problem was discussed in considerable detail in Chapter 3, in connection with solvatochromism. Because of the complexity of the situation, additional information is needed to indicate particular geometries.

The role of the oxidation state of the transition element is particularly clear in the case of iron. This is of considerable practical importance, as glass production technology almost always involves the introduction of a certain amount of iron. Iron(III) ions in their high-spin $d^5(t_{2g}^5)$ configuration are practically colourless, whereas iron(II) in its $d^6(t_{2g}^4 e_g^2)$ configuration causes the occurrence of various colours, especially blue green in calcium-silicate glass. Thus glass can be decolourised by oxidising iron(II) to iron(III), which is sometimes carried out by nitrates, sometimes with a trace of manganese dioxide. We shall discuss doping with iron oxides in more detail below (Section 8.2.2).

8.2.1 Glass doped with Cu^{2+} ions

Glass is doped with oxides, in this case with copper(II) oxide, in high-temperature conditions (ca. 1500 °C) in air. Cu_2O and O_2 are formed under such conditions, which means that the glass contains both Cu(II) and Cu(I) ions. The latter, with their d^{10} electronic configuration, do not manifest split energy levels in the crystal field, and thus do not contribute to the colour of the glass. However, there is the

possibility of contributions from charge-transfer transitions if they occur in the
visible region of the spectrum.

In the case of Cu(II) ions, the issue is considerably more complex, as first of all
there occurs a Jahn-Teller distortion, which results either in a tetragonal splitting of
the energy levels and consequently an increase of the number of absorption bands, or
– more commonly – in significant broadening of the single absorption band expected
for an octahedral d^9 ion. In addition, copper(II) ions have a strong tendency to
variable geometry – the possibilities for four-, five-, or six-coordination has led some
authors to put forward the concept of the 'plasticity' of copper complexes [2]. The
currently prevailing view is that the stereochemistry of copper(II) permits at least five
distinct spatial arrangements of donor atoms. Very extensive spectroscopic data
concerning copper(II) complexes are included in Lever's monograph [3].

The colour of glass doped with Cu(II) ions, and in particular experimental
absorption spectra, do not offer an opportunity of directly observing the situation.
Substantial experimental and interpretative material (including in particular the use
of ESR spectra) can be found in numerous publications; for example, publication [1]
presents the electronic absorption spectra of various glasses doped with Cu^{2+} ions.

As can be seen from Figure 8.1 and Figure 8.2, as well as from data in the relevant
publications, the maxima of absorption spectra differ significantly from one another.
In silicate-lithium glasses, the maximum is located at 740 nm, in sodium glasses at
810 nm, in potassium glasses at 850 nm, and in rubidium glasses at 900 nm. The form
of the band, however, is fundamentally the same and characteristic of Cu(II)
compounds; it is of asymmetrical contour with a substantial half-width.

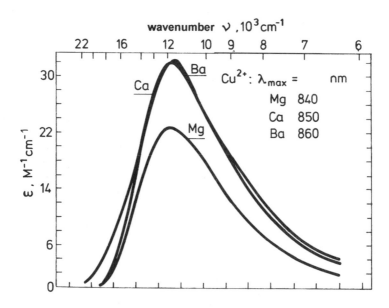

Figure 8.1 Absorption spectra of phosphate glasses $M^{II}O-P_2O_5$, where M = Mg, Ca, Ba, doped with
Cu^{2+} [1].

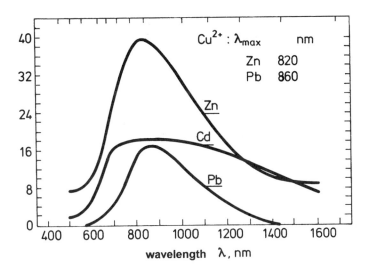

Figure 8.2 Absorption spectra of phosphate glasses $M^{II}O-P_2O_5$, where $M = Zn$, Cd, Pb, doped with Cu^{2+} [1].

This band obviously has less effect on the colour when the absorption maximum occurs above 800 nm, but the large band width means there is still some absorption of visible light. The observed colour is thus chiefly determined by the absorption minimum (transmission maximum), which almost invariably occurs at approximately 400 to 500 nm. As a result the colour is normally within the blue and blue green colour area. The exact hue depends of course on the precise composition of the glass, the temperature, and the various mineral admixtures present.

The blue green colour of glass doped with Cu(II) ions may completely disappear, or shift dramatically, as a result of reductant action. For example, SnO causes the reduction of Cu(II) to Cu(I); as a result, Cu_2O molecules are formed, which are characterised by a red colour due to a charge-transfer transition $O^{2-} \rightarrow Cu^+$. Reduction may also be caused by the presence of carbon in the process of melting the ingredients of the glass. It has been found [4] that in many cases such a situation occurs mostly in glasses with a low content of Na_2O. The proportion Cu^{2+}/Cu^+ also determines the intensity (brightness) of the colour of the glass.

Colourless silicate glass containing Cu(I) may by coloured by oxidation using ultraviolet radiation, which also causes a shift of the absorption maximum from 780 to 815 nm [4]. In studies of the properties and symmetry of Cu^{2+} ions in glass, it is assumed that a broad electronic spectrum corresponds to a tetragonal structure. In that case, three absorption bands should be expected, corresponding to the transitions from term $^2B_{1g}$ to terms $^2A_{1g}$, $^2B_{2g}$, and ^2E_g. Some authors [5] assume that the observed maximum is a transition to the second of the above terms. Kłonkowski [1] showed that the energy of this transition decreases monotonically with the so-called optical basicity of the glass, λ_{cal}, both in crystalline and vitreous systems R_2O-SiO_2 and in aluminium silicate glasses $R_2O-Al_2O_3-SiO_2$ (for $R = Na$, K, Rb, Cs).

The theoretical optical basicity may be calculated from the following formula for silicate glasses with the composition $xR_2O \cdot (100 - x)SiO_2$:

$$\lambda_{cal} = \frac{x}{\gamma_R^{(200-x)}} + \frac{2(100-x)}{2.09(200-x)},$$

where γ_R is a modifying basicity parameter which depends on the kind of glass.

The notion of theoretical optical basicity was introduced and discussed by Duffy and Ingram [6, 7] as well as by Kłonkowski [1].

8.2.2 Glass doped with iron oxides

Iron is one of the most common components of glass, especially in connection with the production of window glass from sand. In sodium-calcium-silicate glasses, this element occurs both in the +2 and in the +3 oxidation state. The effective colour is determined by the proportion of these ions and the intensity of coloration depends on their total content.

The basic issues are related to the different character of the absorption spectra of Fe^{3+} and Fe^{2+} compounds. In the case of Fe^{2+} ions, one must first of all take into account the high-spin configuration and octahedral symmetry. It is to these ions that the absorption band at 1050 nm in the glass under consideration is ascribed. This corresponds well with the position of the band in solution. As can be seen from Figure 8.3 [4], the transmission maximum falls at approximately 500 nm, which causes the occurrence of a green colour.

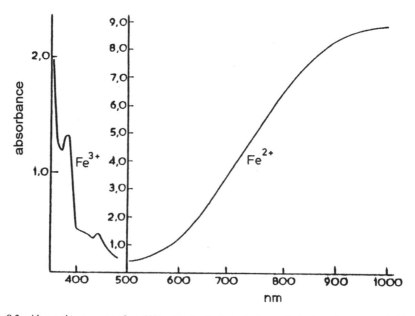

Figure 8.3 Absorption spectra of an SiO_2-Na_2O-CaO (soda-lime-silica) glass doped with Fe^{2+}; Fe^{3+}.

In the case of Fe^{3+} ions with a high-spin configuration and octahedral symmetry, there is only one sextet term ^6S, plus quartet and doublet terms. Since there are no spin-allowed transitions, there should be no colour-generating absorption bands. However, in aqueous solutions, there is a yellow or yellowish red coloration, mostly due to the hydrolysis of Fe(III) salts.

Of course, this situation does not occur in glass; nevertheless weak absorption bands have been observed at 380, 420, and 435 nm [4]. They are spin-forbidden bands and the sextet-quartet transitions show a strongly negative dependence on Dq [3]. As a result, in tetrahedral complexes FeL_4 the wavenumbers of these transitions are higher than in octahedral complexes FeL_6. It has been found, for instance, that for the chromophores $Fe(III)O_6$ and $Fe(III)O_4$, transition energies should be, respectively, 11 000 and 22 000 cm^{-1}.

In aqueous solutions, the following wavenumbers of sextet-quartet transitions have been found [8]: for $^4T_{1g}$ – 12 600, for $^4T_{2g}$ – 18 500, and for $^4A_{1g}, ^4E_g$ – 24 300 and 24 600 cm^{-1} (respectively, 794, 540, 411, and 406 nm). Just as in the case of the $[Fe(H_2O)_6]^{3+}$ ion, where there is a CT transition at 42 000 cm^{-1} (238 nm), in glass a band has been found at 225 nm ascribed to Fe^{3+} ions.

It should be noted that the position of the long-wave band in the infrared, connected with Fe^{2+} ions, depends on the type of glass, and ranges from approximately 950 nm for lithium-silicate glass with the composition $3Li_2O \cdot 7SiO_2$ to 1300 nm for glass with the composition $Li_2O \cdot P_2O_5$. Thus, it is rather the content of Fe^{3+} ions that has a concrete effect on the colour of glass doped with iron compounds, particularly as the positions of the absorption bands of these ions vary greatly depending on the composition of the glass. The colour may change within the blue green and yellow green area when the Fe(II)/Fe(III) ratio changes from 0.5 to 0.3.

8.2.3 Glass doped with manganese oxides

In sodium-calcium-silicate glasses, manganese occurs in the +2 and +3 oxidation states. As is well known, high-spin Mn(II) complexes are essentially colourless, but for large concentrations a pink coloration is observed. In contrast to Mn(II), Mn^{3+} ions cause the occurrence of a purple colour. This is related to the contour of the absorption spectrum of this ion. The spectrum of sodium-calcium-silicate glass is given in Figure 8.4 [4].

As can be seen, strong light transmission occurs both at approximately 375 nm and in the 750–800 nm range, which causes the occurrence of a purple colour (through additive mixing of blue and red). The absorption band at about 490 nm, i.e. in the green range, of course facilitates the appearance of purple.

The absorption bands of Mn^{2+} ions are very characteristic in the case of octahedral symmetry. They show exceptionally small intensities; for instance for the transition $^6A_{1g} \rightarrow {}^4A_{1g}, {}^4E(G)$ in MnF_2, the coefficient ε is 0.15 M$^{-1} \cdot$cm^{-1}. This band is also very narrow. The absorption band with a wavelength of 430 nm (23 250 cm^{-1}) in sodium-calcium-silicate glass is ascribed to the presence of Mn^{2+} ions. In the spectrum of $[Mn(H_2O)_6]^{2+}$ the band with a wavenumber of 23 190 cm^{-1} is identified as the transition $^6A_{1g} \rightarrow {}^4T_{2g}(G)$ [9].

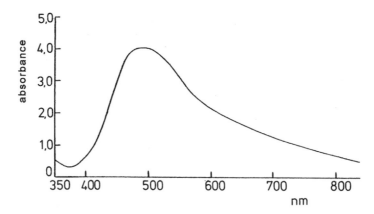

Figure 8.4 Absorption spectrum of a soda-lime-silica glass doped with Mn^{3+}.

The comparison made by Lever [3] can be used to compare the absorption spectra of Mn(III). In the aqua-ion (that is for the chromophore MnO_6), one absorption band is found, with a wavenumber of $21\,000\,cm^{-1}$ (ca. 476 nm). One absorption band, corresponding to the transition $^5E_{2g} \rightarrow {}^5T_{2g}$ may be formally expected for the high-spin d^4 configuration. However, due to the degeneracy of the ground state, Jahn-Teller deformation occurs and as a consequence a tetragonal (or even rhombic) splitting takes place. The splitting may be of different magnitude and sometimes shows up as band asymmetry.

This is precisely the situation that can be observed in the glass spectrum presented in Figure 8.4. In accordance with the energy diagram and interpretation proposed in [3], the observed band corresponds to the transition from $^5B_{1g}$ to one of the split levels of the higher term $^5T_{2g}$, which however is generally represented as $^5B_{1g} \rightarrow {}^5B_{1g}$ (as the splitting often cannot be confirmed experimentally).

The colour of Mn(II) compounds, for large ion concentrations, is one of the examples which indicate that even though there are no spin-allowed transitions in these compounds (for high-spin configurations), spin-forbidden transitions may of course also be the source of colour if they occur in the visible range.

8.2.4   Glass doped with chromium oxides

As is well known, the most stable compounds of chromium occur in the +3 and +6 oxidation states, such as the oxides Cr_2O_3 and CrO_3. The spectroscopic properties and colour of Cr(III) compounds have been discussed several times in this book, and there is a great deal on this subject in the literature.

The spectrum of glass doped with chromium compounds is shown in Figure 8.5 [4], and Figure 8.6 presents the spectrum of the CrO_4^{2-} ion.

The right hand side of Figure 8.5 shows the spectrum of the chromophore $Cr^{III}O_6$; it is important for the colour of the glass. For Cr(VI), an absorption maximum occurs in the ultraviolet at approximately 370 nm, but the arm of this band reaches into the visible range. If one also takes into account the considerable difference in the

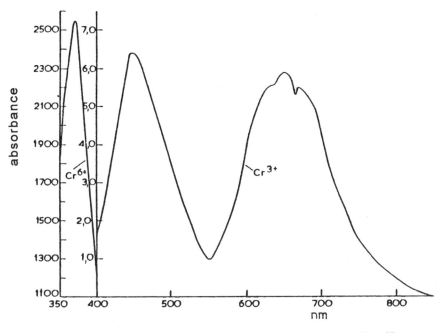

Figure 8.5 Absorption spectra of soda-lime-silica glass doped with Cr^{III}, Cr^{VI}.

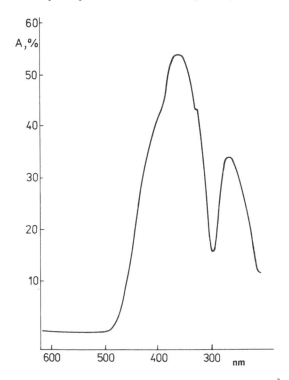

Figure 8.6 Absorption spectrum of the chromate anion, CrO_4^{2-}.

Table 8.3 Parameters of absorption spectra and of Cr(III) crystal field of some glasses (after [11])

Glass	$^2E \leftarrow {}^4A_2$	$^2T_1$	$^4T_2 (10Dq)$	$^4T_1 (F)$	B	C/B
Alumina-phosphate $Al_2O_3 - 28.9$ $P_2O_5 - 71.1$	14760	15630	14500	21500	761	4.08
Calcium–phosphate $CaCO_3 - 33.3$ $P_2O_5 - 67.7$	14770	15630	14750	21740	752	4.15
Soda-phosphate $Na_2CO_3 - 33.3$ $P_2O_5 - 67.7$	14750	15550	15080	22080	745	4.19
Soda-calcium-silicate $SiO_2 - 73.0$ $Na_2CO_3 - 14.0$ $CaCO_3 - 13.0$	14760	15560	15200	22250	770	4.15
Alumina-lithium-borate $B_2O_3 - 65.0$ $Li_2CO_3 - 20.0$ $Al_2O_3 - 13.9$ $As_2O_3 - 1.0$	14710	15600	16420	23530	732	4.29

intensity of light absorption, due to the fact that in the case of Cr(VI) the relevant transition is a CT transition $O^{2-} \rightarrow Cr^{6+}$, its effect on the colour of the glass cannot be disregarded. In industrial practice, the effect of Cr(VI) is eliminated by introducing appropriate reductants during glass-making (e.g. As_2O_3).

The presence of Cr^{3+} in glass in practice determines the occurrence of a green colour, which also means that although a change of the composition of the glass may lead to a change of intensity, both the position of bands and the overall contour will remain virtually unchanged. Some data are included in Table 8.3 [10] and in Figure 8.7 [11].

The doping of glass with Cr^{3+} ions has found an important application in the construction of luminescent solar concentrators (LSC) [12].

On the basis of Table 8.3, it can be concluded that in many glasses the energy of the $^4T_2$ level is lower than that of the ^2E term, which results in emission from that state. This sort of situation is characteristic of the weak crystal field of the d^3 configuration. Thus, it can be expected that all the above-mentioned glasses will be green (or yellow green), as their $10Dq$ values are within the $14\,500–16\,000\,cm^{-1}$ range.

At high partial pressures of oxygen in the technological process, Cr(III) may be oxidised, and the redox equilibrium Cr(III)/Cr(VI) may shift rightwards, which may result in a noticeable colour change. CrO_4^{2-} ions are yellow and the above equilibria generally cause the occurrence of different shades of yellow green colours. It has been found that an increase of the alkalinity of the glass increases the concentration of Cr(VI).

Also important for the colour of the glasses under discussion is the fundamental absorption edge in the ultraviolet region. This value depends on the composition of the glass and, for instance, an increase of the amount of PbO (in lead-silicate glasses)

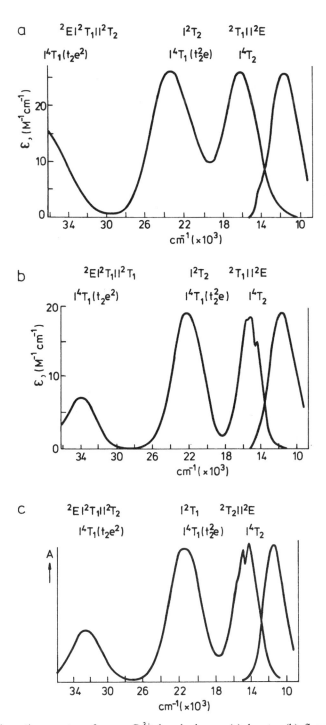

Figure 8.7 Absorption spectra of some Cr^{3+}-doped glasses: (a) borate; (b) fluorophosphate; (c) phosphate [10].

shifts the absorption edge as far as approximately 400 nm, causing the occurrence of colourations from yellow green to brown.

As has been shown [4], under the specific conditions of glass-making, there is a possibility of the occurrence in glass of Cr(IV) and Cr(II) ions. In aluminium-calcium glasses, Cr(IV) causes the occurrence of a blue colour. This may be achieved by slowly oxidising Cr(III) in a nitrate melt with a small quantity of Cr_2O_3 (0.02% by weight). In the absorption spectrum of such glass, there is a band at 610 nm, and an inflection at 790 nm. It is worth noting that the CrO_4^{4-} ion exists; its sodium salt is dark green, while the strontium and barium salts are dark blue.

8.2.5 Glass doped with cobalt oxides

The two principal cobalt oxides are CoO and Co_3O_4. The former is a Co(II) compound, while the composition of the latter is given by the formula $Co_2^{III}Co^{II}O_4$. CoO is olive green; it acts as a pigment on reaction with SiO_2 and Al_2O_3. In Co_3O_4, whose structure is that of a normal spinel, Co^{II} has tetrahedral symmetry, Co^{III} octahedral.

The colours resulting from the doping of glass with cobalt oxides are thus mostly connected with the +2 and +3 oxidation states. Figure 8.8 presents the absorption spectrum of sodium-calcium-silicate glass doped with cobalt oxides.

The colour of Co(II) compounds has been the subject of more detailed discussions elsewhere. Figures 3.8 and 5.10 presented the spectra of $[Co(H_2O)_6]^{2+}$ and of $CoCl_2$ in organic solvents. Important issues of colour are related to the type of coordination. Co(II) coordination compounds with O_h symmetry are usually pink, while in tetrahedral coordination they are blue. These problems have been the subject of many publications. Some issues connected with the use of trichromatic colorimetry and the determination of chromaticity coordinates, the dominant wavelength, and hue angles for $CoCl_2$ in organic solvents were discussed in Chapter 5; they have been set out in reference [13].

On the basis of these data, it can be concluded that blue colour, which is usually observed for glasses with cobalt oxides, is mostly related to the tetrahedral coordination of Co^{2+} ions, i.e. the chromophore CoO_4. This corresponds to

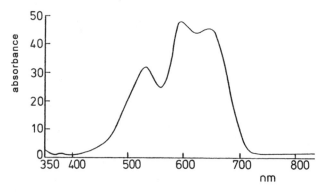

Figure 8.8 Absorption spectrum of soda-lime-silica glass doped with cobalt oxide (cf. Figure 3.8 for $[Co(H_2O)_6]^{2+}$ and Figure 5.12 for Co^{2+} in dimethylformamide) [4].

a pattern of bands at approximately 600 nm. However, there is another, only slightly less intense, band at about 530 nm, which is characteristic of octahedral compounds of Co(II). At the same time, it should be noted that in the spectrum of $[Co(H_2O)_6]^{3+}$, that is for a compound of cobalt in the +3 oxidation state with a low-spin configuration, there are two absorption bands in the visible range, one at about 600 nm, the other at ca. 400 nm. Thus, if it is found that it is rather Co(II) that occurs in glasses doped with cobalt, it should be concluded that a certain portion of Co^{2+} ions also occupy octahedral positions. An additional reason for the intensity of the blue colour is the strong light transmission in the 380–400 nm range (Figure 8.8).

Replacement of Na_2O with potassium oxide causes the occurrence of a blue colour of greater purity, and the spectrum shows a slight bathochromic shift (about 10 nm). The most pronounced changes are observed in borate glasses [4]. A small amount of sodium oxide colours glass pink, and by increasing the amount of Na_2O, the colour is changed gradually to pink violet and blue.

Although Co^{2+} ions are assumed to be the main form of cobalt in glass, there are also data in the literature concerning Co^{3+} (e.g. [14]). The authors of that paper found that such ions occur in sodium-silicate glasses containing more than 50% of Na_2O. If the amount of this oxide is increased, the colour of the glass changes from blue, through blue green, to yellow green and yellow.

8.2.6 Glass doped with nickel oxides

In these glasses the principal oxide is NiO, introduced as a green powder. In glass one should take into account the occurrence of Ni^{2+} ions, in both octahedral and tetrahedral symmetries. For a particular equilibrium of these ions, the colour of glass depends not only on the stability constant, but also on the particular composition of the glass. In alkaline-calcium-silicate glasses with Li_2O, Ni(II) shows octahedral symmetry; with K_2O, tetrahedral; and with Na_2O, an intermediate situation. Accordingly, lithium glasses are light yellow, potassium glasses are purple, and sodium glasses are an intermediate brown colour.

Figure 8.9 shows the absorption spectrum of sodium-calcium-silicate glass doped with nickel oxide. The proposed interpretation [4] of this spectrum does not appear to be unequivocal. The author assumes that the bands at 450 (main band), 930, and 1800 nm correspond to transitions in O_h symmetry, while those at 560, 630, and 1200 nm correspond to transitions in T_d symmetry.

For the sake of comparison, we shall consider some spectroscopic data for these two types of symmetry in Ni(II) compounds. According to Lever [3], undistorted octahedral complexes show 3 transitions, in the following wavenumber ranges: $7000–13\,000\,cm^{-1}$ (1400–770 nm), $11\,000–20\,000\,cm^{-1}$ (900–500 nm), and $19\,000–27\,000\,cm^{-1}$ (520–370 nm). For the $[Ni(H_2O)_6]^{2+}$ ion, the positions of the three bands are as follows: 1176, 724, and 395 nm. It should be noted that the NiO/MgO system shows almost identical band positions. However it should be stressed that transitions of wavelengths as long as 1800 nm have not so far been observed in spectra of octahedral nickel(II) complexes. The 1800 nm band may therefore perhaps be more convincingly assigned to the tetrahedral NiO_4 chromophore – the longest wavelength transition for $NiCl_4^{2-}$ is at 1526 nm.

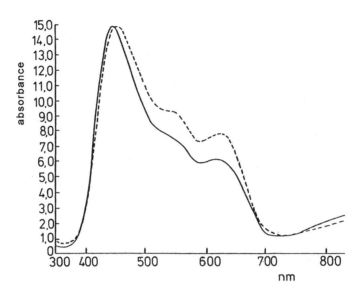

Figure 8.9 Absorption spectra of soda-lime-silica glass doped with nickel oxide (——— annealed; - - - - toughened (quenched)) [4].

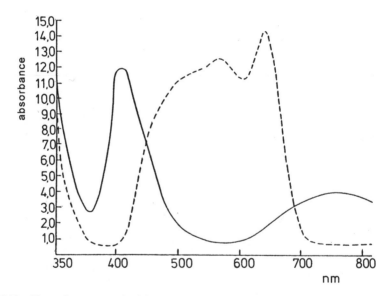

Figure 8.10 Absorption spectra of Ni-doped glasses: ——— sodium borate; - - - - potassium silicate [4].

The greatest similarity to the spectrum of the aqua-ion is revealed by the spectrum of sodium-borate glass with a perfectly formed band at approximately 400 nm (greatest intensity), with a deep minimum at about 550 nm, which usually gives the compounds of nickel(II) a green colour. Potassium-silicate glass shows a completely different absorption spectrum (Figure 8.10).

For borate glass a green colour can be expected, and for potassium-silicate glass a purple colour, owing to strong transmittance in the 700 to 800 and around 380 to 400 nm ranges, and additive colour mixing.

REFERENCES

[1]   Kłonkowski, A., 1987, *Natura wiązania chemicznego i struktura niektórych szkieł tlenkowych*. Zeszyty Naukowe Politechniki Gdańskiej, Chemia, 29, Gdańsk.

[2]   Gažo, J., Bersuker, J.B., Garaj, J., Kabesova, M., Kohout, J., Langfelderova, H., Melnik, M., Senator, M. and Valach, F., 1976, *Coord. Chem. Rev.*, 19, 253.

[3]   Lever, A.B.P., 1984, *Inorganic Electronic Spectroscopy*, 2nd edn., Elsevier, Amsterdam.

[4]   Bamford, C.R., 1977, *Colour Generation and Control in Glass*, Elsevier, Amsterdam.

[5]   Bjerrum, J., Ballhausen, C.J. and Jørgensen, C.K., 1954, *Acta Chem. Scand.*, 8, 1275.

[6]   Duffy, J.A. and Ingram, M.D., 1976, *J. Non-Cryst. Solids*, 21, 373.

[7]   Duffy, J.A. and Ingram, M.D., 1975, *J. Inorg. Nucl. Chem.*, 37, 1203.

[8]   Jørgensen, C.K, 1962, *Absorption Spectra and Chemical Bonding in Complexes*, Pergamon, Oxford.

[9]   Bartecki, A., 1987, *Chemia pierwiastków przejściowych*, WNT, Warsaw.

[10] Andrews, L.J., Lempicki, A. and McCollum, B.C., 1981, *J. Chem. Phys.*, 74, 5526.

[11] Nassau, K., 1983, *The Physics and Chemistry of Color*, New York: Wiley.

[12] Reisfeld, R. and Jørgensen, C.K. 1982, *Struct. Bonding*, 49, 1.

[13] Bartecki, A., Tłaczała, T. and Raczko, M., 1991, *Spectroscopy Lett.*, 24, 559.

[14] Deitzel, A. and Coenen, M., 1961, *Glas. Ber.*, 34, 49.

9. COLOUR IN TEACHING INORGANIC CHEMISTRY

The colour of chemical compounds plays an extremely important role in teaching chemistry, especially descriptive chemistry. This applies not only to transition metal chemistry, but to most elements and compounds. In chemical education curricula, presentations and demonstrations are an important teaching component. These very often rely on the colours of particular compounds, or, for even more effective appeal to the audience, on the appearance or change of colour in the course of the demonstration.

Of course, all the aspects of the occurrence of colour in chemical systems which have been discussed in the preceding chapters (especially in Chapters 3, 4, and 5) may be used in teaching chemistry, but some topics are of importance only to more advanced students and researchers. Until fairly recently chemistry textbooks were printed simply in black and white. Then came occasional, and soon much more frequent, use of colour, both in illustrations and in text, in general chemistry texts, especially for the American market. In recent years a number of general and inorganic textbooks have appeared which are profusely illustrated with coloured photographs and diagrams, though a number of the standard popular under-graduate textbooks continue to eschew colour. Presumably the continuance of black-and-white may be attributed to the need or desire to keep costs at a minimum, especially for new editions of established texts. Thus the recent (1998) second edition of the excellent undergraduate textbook by Greenwood and Earnshaw [1] maintains the high standard of diagrams and Figures which characterised the first edition, but contains no colour. Likewise the fifth edition of the ever-popular Cotton and Wilkinson [2] has no coloured material – it will be interesting to see whether any colour creeps into the long-promised sixth edition, or into the imminent third edition of the (undeservedly) popular text by Shriver, Atkins, and Langford [3]*.

Excellent examples of the use and value of colour in teaching texts can, however, be found in a number of recent textbooks. There are two main aspects, on the one hand there is the use of colour to clarify the text and improve its appearance, on the

---

*Note added in proof: The third edition comes with a free CD-ROM, but with no colour beyond that on the cover.

211

other the use of colour photographs as illustrations. Such photographs may show the colours of the compounds discussed and changes in colour in reactions, both preparative and analytical. Many also include striking colour photographs of chemistry in use or action in 'real-life' situations and applications. The new book by Housecroft and Constable [4] makes much use of colour in text and layout. It contains rather few colour photographs, but these do include an excellent picture of green beryl. Several other textbooks are very generously provided with illustrations in colour, though just occasionally they also illustrate the difficulties in printing colours correctly! In Box 9.1 we give a number of examples taken from one of the better earlier books in this class [5], while in Box 9.2 [6, 7] we give examples from two more recent books from the same stable. Whereas these two boxes list examples in order of appearance, in Boxes 9.3 and 9.4 we list a selection of colour photographs organised into categories from two other profusely illustrated general chemistry books [8, 9]. Although student texts for the American market have, perforce, similar contents, the emphasis on topics illustrated and the level of treatment do vary significantly between authors. We conclude this paragraph by citing two very recent (1997 and 1998) new editions of examples of this genre [10].

Box 9.1

Colour illustrations of particular relevance to chemistry teaching appearing in J.E. Brady and J.R. Holum, *Fundamentals of Chemistry*, 3rd edn., Wiley, New York, 1988:

p. 475	$- KMnO_4$	– violet solution added to pale blue green solution of Fe^{2+}; purple colour disappears
p. 920	$- [Cr(H_2O)_6]^{3+}$	– 3 aqueous solutions – violet
	$- [Cr(H_2O)_3(OH)_3]$ alkaline	– blue + precipitate
	$- [Cr(H_2O)_2(OH)_4]$ strongly alkaline	– green
	$- KCr(SO_4)_2 \cdot 12H_2O$	– solid phase/aq soln. – violet
p. 921	$- CrO_3$	– dark orange red solid
Figure 22.6	$- Cr_2O_7{}^{2-}$	– aqueous solution, reddish orange
	$- CrO_4{}^{2-}$	– aqueous solution, yellow
	$- Cr_2O_3$	– powder, green grey
p. 925	$- Fe^{2+}$	– pale blue green solution
	$- Fe^{3+}$	– yellow solution due to hydrolysis
Figure 22.7		
(left)	$- Fe_4[Fe(CN)_6]_3$	– dark Prussian blue as a result of reaction of $[Fe(CN)_6]^{4-} + Fe^{3+}$ aq
(right)		– architect over blueprint covered with blue pigment
Figure 22.18	– cobalt(III) complexes	– aqueous solutions – 6 different ligands give red, yellow, green, violet, red, violet colours

Box 9.2

Some colour illustrations of particular relevance to chemistry teaching appearing in J.E. Brady, *General Chemistry: Principles and Structure*, 5th edn., Wiley, New York, 1990:

p. 75   – precipitation of yellow lead chromate
p. 98, 145   – colours characteristic of chromium (VI) and manganese (VII)
p. 195   – continuous spectrum; sodium and hydrogen emission spectra
p. 557   – acid/base indicators
p. 580/1   – electrolysis of NaBr, $CuBr_2$ solutions
p. 665   – colour wheel
p. 670   – flame test colours
p. 718   – range of colours for cobalt(III) complexes in solution
p. 804   – emerald and ultramarine

Some of the colour pictures mentioned above, and in Box 9.1, also appear in J.E. Brady and J.R. Holum, *Chemistry: The Study of Matter and Its Changes*, 2nd edn., Wiley, New York, 1996. This book contains further relevant pictures.

Box 9.3

Colour illustrations of particular relevance to chemistry teaching, and to the subject matter of the present monograph, appearing in J.C. Kotz and K.F. Purcell, *Chemistry and Chemical Reactivity*, Saunders, Philadelphia, 1987. Page numbers are quoted in all cases.

Solution chemistry		Transition metal species (solutions)	
of lead	799	aqua-ions	976
of iron	987	chromium species	947
of nickel	989	cobalt complexes	970
Minerals and gemstones		Solid compounds	
amethyst	48,802	copper compounds	88
azurite	8,954	mercury (HgI, HgO, HgS)	99,110
gold	26	sulphides, hydroxides	88
malachite	954	*d*-block complexes	959
orpiment	819	*d*-block compounds	57
realgar	819	Indicators	
ruby	790	acid/base colour changes	566–570
Emission spectra	228	red cabbage	133

This book is also noteworthy for the use of colour to improve readability and comprehensibility. A particularly good example occurs on page 317, where electronegativity trends across and down the Periodic Table are represented by the colours of the rainbow.

Box 9.4

Colour illustrations of particular relevance to subjects covered in the present monograph which appear in H.F. Holtzclaw, W.R. Robinson, and J.D. Odom, *General Chemistry*, 9th edn., D.C. Heath, Lexington, Mass., 1991. Page numbers are quoted in all cases.

Minerals and gemstones		Transition metal species	
amethyst	396	aqua-ions	834,867*
cinnabar	903	chromium in solution	846
copper	862	chromium (VI) solids	880
emerald	811	copper complexes (soln)	854
gold	862		
hematite	69,862	Thermochromism – Cr^{3+} aq	881
malachite	833		
ruby	396	Acid/base indicators	
sapphire	396	standard set of nineteen	567
sulphur	326,696	universal indicator	520
topaz	396		
wulfenite	3	Emission spectra	
		Na, H, Ca, Hg	138
Elements			
selection of	158–9	Fireworks: Sr-coloured	405
tin pest	905		
Lead oxides	908		

* Unfortunately the solution claimed to contain Fe^{3+}aq is yellow, which is the colour of hydroxo-iron(III) species. The true colour of Fe^{3+}aq is pale purple, as in ferric alum crystals, though aqueous solutions would appear colourless at normal concentrations.

Although there are now a number of books making good use of colour in the teaching of inorganic chemistry it is rare to find a discussion of the phenomenon of colour. Sadly, one textbook which does include such a discussion only includes one colour picture of transition metal species, and that in rather inaccurate colour reproduction [11].

There are many publications – books, journals, and articles – devoted to demonstrations of the role and importance of colour in chemical education. The main literature source is the *Journal of Chemical Education*, from which we shall quote just a few examples. Firstly we cite an article [12] in which the author describes seminar classes devoted to the colours of transition metal complexes, where students discuss the theoretical foundations and then demonstrate selected reactions. Table 9.1 lists the topics covered.

The use of colour in teaching chemistry is particularly relevant to explaining and demonstrating the role and use of the Crystal Field (Ligand Field) model in transition metal chemistry. This may be done even at secondary school level. It is a

Table 9.1 Titles of student presentations of colours of transition metal complexes [12]

No.	Title
1.	Patriotic colours of Co(II) in organic solvents
2.	A rainbow array of nickel complexes
3.	Chloro- and thiocyanato-complexes of cobalt(III)
4.	Red, white, blue iron(II) (III) complexes
5.	Precipitates and complexes of Ag(I)
6.	Prussian blue
7.	Precipitates and complexes of iron(II)
8.	Iodocomplexes of mercury
9.	Bromocomplexes of copper
10.	Precipitates and complexes of cobalt(II)

trivial, but none the less centrally relevant, observation that the most salient characteristic of transition metal compounds and complexes is their colour, most commonly due to ligand field transitions. Once this issue has been been understood and become familiar, the perspective can be broadened to include other sources of colour, such as charge transfer transitions. One very commonly encountered example of colour reactions concerns manganese, a particularly appropriate element in view of the large number of oxidation states, most of them fairly stable and most of them coloured (Scheme 9.1 [13]), it exhibits. Intensely coloured compounds exist for oxidation states +7, +6, +5, and +4; conditions for the formation of various coloured complexes of Mn^{3+} have yet to be fully specified. Jeżowska-Trzebiatowska and Bartecki [14], in examining the $KMnO_4$–$SnCl_2$–acetone redox system, demonstrated the existence of an Mn^{3+} complex and reported its absorption spectrum. The colour of this Mn^{3+} complex was, under the conditions used in this investigation, reddish-brown. Manganese(II) compounds are usually only very feebly

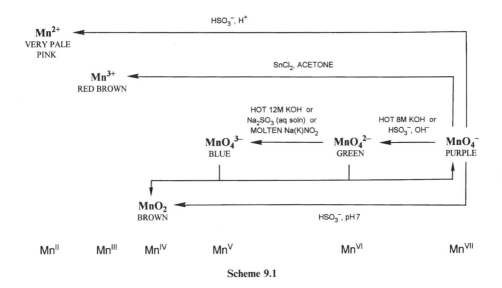

Scheme 9.1

coloured – Mn^{2+} aq is very pale pink – since their *d-d* transitions are, as mentioned earlier in this book, doubly forbidden.

One of the reactions best suited to demonstration lectures is the system bromate (as oxidant) – organic acid (e.g. malonic; as reductant) – catalyst (usually Mn^{2+} or Ce^{3+}); often with $[Fe(phen)_3]^{2+}$ added as indicator [15]. In systems of this kind, oscillatory reactions [16] take place, which result in periodic changes in the concentrations of catalyst, and of the indicator.[1] These oscillations are most strikingly revealed in the appearance and disappearance of the intense orange colour of the $[Fe(phen)_3]^{2+}$ indicator. In the absence of the iron complex, the oscillatory behaviour is still apparent from the alternation between colourless Mn^{2+} and coloured Mn^{3+}, or colourless Ce^{3+} and yellow Ce^{4+}. Figures 9.1 [17] and 9.2 [18] depict the progress of several oscillating reactions.

Special cases of this type of reaction are *clock reactions*, wherein a sudden appearance of an intense colour follows an induction period. Such systems include, for instance, hydrogen peroxide oxidation of iodine. Oscillatory reactions and clock reactions are included in the class of *exocharmic reactions* [19], which are reactions which attract the attention of audiences and give them a certain pleasure. There is no doubt that periodic colour changes, or the swift appearance of a bright colour a short time after mixing colourless solutions, deserve to be described as exocharmic, since they are particularly attractive to audiences and thus are good ambassadors for chemistry.[2] A recent (1996) book entitled 'Chemical Curiosities' [20] contains a large number of experiments of an exocharmic character, many of which involve colour or colour changes, and is generously illustrated with high quality colour photographs. In fact this book has in places a more serious intent than implied by its title, for it includes a plate documenting the solvatochromic behaviour of $[Fe(phen)_2(CN)_2]$ (as has already been mentioned) and also several photographs documenting aspects of transition metal chemistry which appear in all undergraduate courses, including

- reduction of vanadate;

- reduction of permanganate;

- nickel(II) complexes in solution;

- cobalt(II) complexes in solution.

In analytical and descriptive chemistry, use is sometimes made of the fact that a flame can be coloured by the volatile salts of sodium, potassium, calcium, strontium, and barium. Under suitable experimental conditions, these elements give the flame the following colours: yellow, violet, brick-red, carmine-red, and yellow green (as is well known, this phenomenon is used in the production of fireworks). The colour of the flame is connected with the process of metal ion excitation and light emission. The wavelengths of the emitted light are: in the case of sodium, 589 nm; for

---

[1] The parent system is the so-called Belousov-Zhabotinsky reaction.
[2] The importance of colour in attracting school audiences to chemistry has long been recognised – see, e.g. W.D. Wright, *The Rays are not Coloured*, Adam Hilger, London, 1967, Chapter 5.

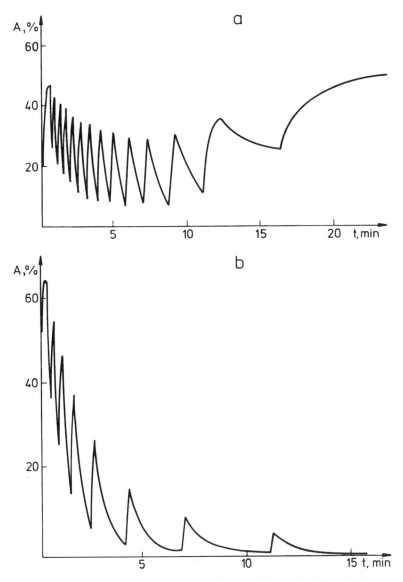

Figure 9.1 Oscillatory curves of the system $CH_2(CO_2H)_2-KBrO_3-MnSO_4-H_2SO_4$ at two concentration ratios $KBrO_3$: malonic acid, (a) >1; (b) <1.

potassium, 405 and 767 nm; for calcium 423, 559, and 616 nm. Strontium emits four lines, and barium five (in the visible range).

An important matter, which has so far not been given enough attention, is the problem of unique descriptions of the colour of a given compound. As has been mentioned above, in some languages, there are more names than in others, or some authors use names which differ from customary descriptions.

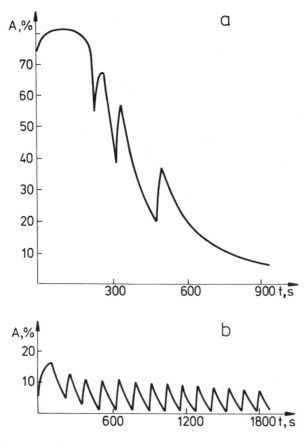

Figure 9.2   Oscillatory curves for the system $HO_2CCH_2CHOHCO_2H-KBrO_3-MnSO_4-H_2SO_4$ at the following initial concentrations $(mol\,dm^{-3})$ of the components:

	(a)	(b)
malic acid	0.1014	0.1544
potassium bromate	0.050	0.063
manganese sulphate	0.066	0.0012
sulphuric acid	2.76	1.25

It should first of all be realised that the unification of colour terminology is practically impossible owing to the immense number of colour hues (colour sensations), in the order of approximately 10^6–10^7. Thus, only numerical indicators characterising the quantity of colour and obtained within a particular method can be understood precisely by a reader/listener. Such methods include the CIE system and the modified systems, CIELAB and CIELUV, as well as the Munsell systems, all of which have been discussed in this book.

A proposal for the systematic classification of the colours of coordination compounds of the transition metals was put forward and discussed a few years ago [21, 22]. Table 9.2 presents a selection from the examples given in these publications.

Table 9.2 Colour classification and systematic colour names of some coordination compounds [21, 22]

Compound[a]	Reported colour	Sample treatment[b]	Colour notation MUNSELL[c]		Colour name ISCC-NBS[d]
$Ni(py)_2Cl_2$	pale yellow green	P	5.5GY	8.5/3	light yellow green
$K_2[PdCl_4]$	yellow greenish brown	G_3	4.5Y	5.5	light olive
$K_3[Fe(CN)_6]$	red	P	3YR	6.5/15	vivid orange
$[Co(NH_3)_5Cl]Cl_2$	dark red violet	P	2.5RP	3/9	deep reddish purple
$[Ni(en)_3]Cl_2.2H_2O$	orchid	P	4.5P	4.5/14.5	vivid purple
$Ni(en)_2Cl_2$	blue	P	6PB	6/9.5	brilliant blue
$Ni(py)_4Cl_2$	pale blue	G_1	6.5B	8/2	very pale blue

[a] en = ethane-1,2-diamine; py = pyridine. [b] P = assessed as prepared; G = assessed after grinding, G_n representing particle size from coarse (G_1) to very finely ground (G_4). [c] The revised Munsell notation of S.M. Newhall, D. Nickerson, and D.B. Judd, *J. Opt. Soc. Am.*, **33**, 385 (1943) – the three symbols indicate hue, value, and chroma. [d] K.L. Kelly and D.B. Judd, *The ISCC-NBS Method of Designating Colors and a Dictionary of Color Names*, NBS Circular 553, National Bureau of Standards, Washington, D.C., 1955. These ISCC-NBS colour names differ significantly from those given in, e.g., A. Kornerup and J.H. Wanscher, *The Methuen Book of Colour*, 2nd edn. and 3rd edn., Eyre Methuen, London, 1969 and 1978 (see Table 1 of reference [22]).

It is evident how difficult it is to give unique and universally accepted descriptions of colours on the basis of the existing classification systems, such as the Munsell system or ISCC-NBS, especially in comparison with the perceived colour (given in column 2).

In some cases, the differences are striking. For instance, the compound $[Co(en)_3]Cl_3.3H_2O$ was characterised as dark orange when it was first obtained; according to ISCC-NBS it is bright orange yellow, and according to Munsell it is described by the symbol 7.5 YR.

It is worth noting how the colour of a given compound changes as a result of finer or coarser comminution. A good example is provided by $K_3[Fe(CN)_6]$, whose colour is generally described as ruby red. As a result of finer and finer comminution, not only do the Munsell shades change, but also the colours described according to the ISCC-NBS scheme change as follows: bright orange yellow, bright yellow, light yellow.

Although the initiative of the authors of the above proposal is commendable, it does not seem that their system is ideal. However, efforts in this direction should be continued, especially with attempts to incorporate the methods of trichromatic colorimetry. One would particularly like to see an improved system for the characterisation of coloured solutions of transition metal complexes.

REFERENCES

[1] Greenwood, N.N. and Earnshaw, A., 1998, *Chemistry of the Elements*, 2nd edn., Pergamon, Oxford.

[2] Cotton, F.A. and Wilkinson, G., 1988, *Advanced Inorganic Chemistry*, 5th edn., Wiley, New York.

[3] Shriver, D.F., Atkins, P.W. and Langford, C.H., 1994, *Inorganic Chemistry*, 2nd edn., Oxford University Press.

[4]   Housecroft, C.E. and Constable, E.C., 1997, *Chemistry – An Integrated Approach*, Addison-Wesley-Longman, Harlow.

[5]   Brady, J.E. and Holum, J.R., 1988, *Fundamentals of Chemistry*, 3rd edn., New York: Wiley.

[6]   Brady, J.E., 1990, *General Chemistry: Principles and Structure*, 5th edn., New York: Wiley.

[7]   Brady, J.E. and Holum, J.R., 1996, *Chemistry: The Study of Matter and Its Changes*, 2nd edn., New York: Wiley.

[8]   Kotz, J.C. and Purcell, K.F., 1987, *Chemistry and Chemical Reactivity*, Saunders, Philadelphia.

[9]   Holtzclaw, H.F., Robinson, W.R. and Odom, J.D., 1991, *General Chemistry*, 9th edn., D.C. Heath, Lexington, Mass.

[10] Atkins, P. and Jones, L., 1997, *Chemistry – Molecules, Matter, and Change*, 3rd edn., Freeman, W.H., New York; McMurray, J. and Fay, R.C., 1998, *Chemistry*, 2nd edn., New Jersey: Prentice Hall.

[11] Kotz, J.C. and Treichel, P., 1996, *Chemistry and Chemical Reactivity*, 3rd edn., Saunders College Publishing, Philadelphia.

[12] Rodgers, G.E., 1988, *J. Chem. Educ.*, **65**, 543.

[13] Pearson, R.S., 1988, *J. Chem. Educ.*, **65**, 451.

[14] Jeżowska-Trzebiatowska, B., Bartecki, A. and Chmielowska, M., 1959, *Bull. Acad. Pol. Sci. Ser. Chim.*, 7, 485.

[15] Cooke, D.O., 1979, *Inorganic Reaction Mechanisms*, The Chemical Society, London, Chapter 7.

[16] Scott, S.K., 1994, *Oscillations, Waves, and Chaos in Chemical Kinetics*, OUP, Oxford.

[17] Bartecki, A. and Tłaczała, T., 1978, *Kinet. Katal.*, **20**, 43.

[18] Tłaczała, T., Ph.D. thesis, Wrocław, 1978.

[19] Ramette, R.W., 1980, *J. Chem. Educ.*, **57**, 68.

[20] Roesky, H.W. and Möckel, K., 1996, *Chemical Curiosities*, VCH, Weinheim.

[21] Wimmer, F.L. and Poncini, L., 1984, *Talanta*, **31**, 651.

[22] Wimmer, F.L. and Poncini, L., 1985, *Acta Chim. Hung.*, **120**, 235.

CONCLUDING REMARKS

As stated in the Preface, the idea of a book devoted to quantitative aspects of the colour of chemical compounds, particularly of metals, was conceived several years ago. In the meantime it has become the natural consequence of many years of studying the absorption spectra of such compounds. Over the course of nine chapters the authors have endeavoured to present primarily the formal aspects of the purposefulness and practicability of the definition and use of colour, particularly with respect to transition metal compounds and complexes. More importantly, the underlying objective has been to impart to the reader a fascination with colour, a fascination which the authors themselves have found to increase steadily as work on this monograph has progressed.

A chemist usually pays little attention to the question of colour, even though this property confronts him or her from the moment of the first encounter with chemistry, and frequently thereafter, in lecture demonstrations and, particularly, in laboratory classes. The dependence of perceived colour on many factors, which is a problem we have discussed to a certain extent in this book, means that colour appears to be a somewhat capricious quality, and therefore considered not to be a fully trustworthy parameter or source of information. However, if experimental conditions are strictly controlled, colour – and especially its quantitative expression in terms of, for example, the CIE parameters – can be an important source of analytical and structural information. The authors have endeavoured to demonstrate this in numerous experimental systems and situations, as well as through computer simulation of absorption spectra and colour.

The quantitative measurement of colour can be satisfactorily carried out, particularly for solutions. Here several problems, which usually accompany measurements of reflectance spectra and colour of solid substances, disappear. Some aspects have been discussed in this book, but it is a vast subject. Such measurements can be used to monitor various equilibrium and kinetic aspects involving coordination compounds, to study the role of solvents and solvation (solvatochromism), to detect unstable complexes, and indeed to afford insight into a number of other properties – all this especially for complexes of the *d*-block transition metals, lanthanides, and actinides. One of the most important tasks for the future is the unique characterisation of experimental conditions, in other words standardisation of the quantitative measurement of colour. Only then will the relevant parameters give unequivocal information about the colour of a given substance.

To conclude, the authors would like to express their hope that this book has introduced readers to the problems inherent in the colour of chemical compounds, and perhaps may prompt them to give these issues some thought and to develop and refine their own approach to this field.

INDEX

We have tried to index chemical substances – compounds, minerals, pigments, gemstones, and constituents of glasses – comprehensively. Index entries for topics and parameters which are mentioned very frequently – such as CIE systems, chromaticity, ligand field, *d-d* transitions, *Racah B*, and charge-transfer – give only the pages of their first mention and definition, and of subsequent important appearances. We have not provided an Author Index, but people's names are indexed when commonly used in identifying equations, formulae, or parameters. Index entries in the sequence C1 to C15 are Figure numbers for the colour plates.

223